ADSORPTIVE IRON REMOVAL FROM GROUNDWATER

Adsorptive Iron Removal from Groundwater

DISSERTATION
Submitted in fulfilment of the requirements of
the Academic Board of Wageningen University and the Academic Board of the
International Institute for Infrastructural, Hydraulic and Environmental
Engineering
for the Degree of DOCTOR
to be defended in public
on Wednesday, 19 December 2001 at 13:00 h in Delft, The Netherlands

by

SAROJ KUMAR SHARMA
born in Kathmandu, Nepal

This dissertation has been approved by the promoter:
Prof. dr. ir. J.C. Schippers, IHE Delft/Wageningen University, The Netherlands

Members of the Awarding Committee:
Prof. dr. W. Rulkens, Wageningen University, The Netherlands
Prof. dr. V.A. Snoeyink, University of Illinois, USA
Prof. dr. J.W. Geus, Utrecht University, The Netherlands
Prof. dr. ing. R. Gimbel, University of Duisburg, Germany
Prof. ir. J.C. van Dijk, Delft University of Technology, The Netherlands
Prof. dr. ir. G.J.F.R. Alaerts, IHE Delft/Delft University of Technology, The Netherlands

ISBN 90 5410 430 9

Contents

Acknowledgements

Many people have contributed their time, energy, ideas, experience and encouragement to help me complete this study, for which I would like to take this opportunity to express my appreciation.

First and foremost, I extend my thanks to my supervisor Prof. Dr. ir. J. C. Schippers and my mentor Dr. ir. B. Petrusevski who were both instrumental in providing focus and channelling the study in a meaningful direction, and for providing valuable feedback at the critical junctures. Prof. Schippers was always inspiring and educating. I enjoyed working with him and learned a lot from his broad knowledge and vast experience. Dr. Petrusevski was very friendly, encouraging and always a source of help when needed.

A special gratitude to Dr. M. R. Greetham for his encouragement and guidance during the difficult initial period of this study. I also owe Ir. A. N. van Breemen and Dr. M. D. Kennedy my sincere thanks for their help and advice from time to time.

Gratefully acknowledged is the financial support of Kiwa N.V. Water Research and five Water Supply Companies in The Netherlands, namely: Water Supply North West Brabant (WNWB), Water Supply East Brabant (WOB), Water Supply Gelderland (WG), Water Supply Drenthe (WMD) and Water Supply Overijssel (WMO) to conduct this study.

Ir. M. Groenendijk and Ing. M. van der Haar of WNWB were very helpful in providing all necessary support during the pilot plant study and making available filter media from different plants from time to time. I would also like to thank Dr. G. IJpelaar, Dr. ir. B. Heijman, Ir. J. Kappelhof and Ir. G. Reijnen from Kiwa N.V. Water Research and Dr. E. Orlandini who were always encouraging and willing to answer questions and pass on what they had learned.

Prof. Dr. J. W. Geus and Ing. C. van Bennekom are acknowledged with thanks for their valuable suggestions as members of the Research Project Steering Committee. I am also thankful to all the members of the Contact Group Groundwater Filtration-Kiwa for their constructive comments and guidance during the course of this study. I am grateful to Ms. W. Sturrock for the proof-reading and English corrections of the draft of this thesis and to Ir. A. N. van Breemen for the Dutch translation of the abstract.

Thanks are also due to seven M.Sc. students involved in this study namely Mr. D.O. Owore, Mr. H. Mutikanga, Mr. R.A.B.S. Mendis, Mr. H. Karunatilake, Mr. C. Sebwato, Mr. D.K. Kironde and Mrs. N. Uyouko for their hardwork and valuable contribution.

My sincere thanks and appreciation to the IHE laboratory staff namely Mr. F. Kruis, Mr. F. Wiegman, Mr. C. Bik, and Mr. P. Heerings for their help and experimental support. I am also thankful to IHE staff members for their support during the study. Thanks to my colleagues Ingrida Bremere and Mustafa Mousa and all the Nepalese participants at IHE in the last six years for their direct and indirect help, encouragement and for providing good company, making my stay in Delft memorable.

Finally, I would like to thank my mother, my wife Sushmita and our son Sumeet for their continuous support, endless patience and understanding.

Abstract

Iron is commonly present in groundwater worldwide. The presence of iron in the drinking water supply is not harmful to human health, however it is undesirable. Bad taste, discoloration, staining, deposition in the distribution system leading to aftergrowth, and incidences of high turbidity are some of the aesthetic and operational problems associated with iron in water supplies. Iron removal from groundwater is, therefore, a major concern for water supply companies using groundwater.

Aeration followed by rapid sand filtration is the most commonly used method for the removal of iron from groundwater. Different mechanisms (physical, chemical, and biological) may contribute to the removal of iron in filters and the dominant mechanism depends on water quality and process conditions applied. Under the commonly applied treatment conditions in iron removal plants, the oxidation-floc formation mechanism (floc filtration) is commonly believed to be dominant. In this mechanism soluble iron(II) present in anoxic groundwater is oxidised to insoluble iron(III) and after precipitation, iron hydroxide flocs are removed in the rapid sand filters. The second mechanism, adsorption-oxidation (adsorptive filtration) may play a role as well and has several potential advantages over the oxidation-floc formation mechanism, namely longer filter run, better filtrate quality, shorter filter ripening time, and less backwash water use and sludge production. In the adsorption-oxidation mechanism, the iron(II) is removed by adsorption onto the surface of the filter media. Subsequently, in the presence of oxygen, the adsorbed iron(II) is oxidised forming a new surface for adsorption. In conventional iron removal filters, the adsorption-oxidation mechanism is expected to be responsible for the removal of an important part of iron entering the filter bed in iron(II) form. Adsorptive filtration is most likely the dominant mechanism in dry filters and in sub-surface iron removal.

Water supply companies are continually seeking means to improve the process efficiency of iron removal from groundwater in order to minimise the deposition of iron in distribution networks, the backwash water use, and the volume of sludge produced. The WHO recommended guideline value of iron in drinking water is 0.3 mg/l and the EC directive has set a parametric value of 0.2 mg/l. In the Netherlands, several water supply companies are aiming at an iron concentration ≤0.03 mg/l in water supplies. Meeting these stringent requirements of iron in the water supply and backwash water treatment in developed countries, and reducing the operation and maintenance costs of distribution systems worldwide will require a more efficient removal and/or minimisation of the iron currently passing through the filters. Until now, it has been generally believed that, regardless of the water quality, the treatment approach was based on physical removal of the iron hydroxide flocs. However, depending upon the water quality and process conditions applied, the application of adsorptive filtration may result in a higher process efficiency than floc filtration. A better understanding of the different mechanisms involved in the iron removal process is, therefore, essential to optimise the design and operation of iron removal filters in terms of run time, filtrate quality, and overall treatment costs. Despite several advantages, the adsorption-oxidation mechanism has not been knowingly employed as the sole or dominant mechanism of iron removal in wet filters yet.

The goal of this research has been to examine the potential of adsorptive iron removal as an alternative to the conventional floc formation method and to investigate the factors governing the adsorptive iron removal process, particularly the mechanisms involved under anoxic conditions. This was accomplished by i) measuring the iron(II) adsorption capacities of several new filter media, iron oxide coated sand from iron removal plants, and iron hydroxides, ii) analysing the effect of water quality parameters on adsorption capacity, iii) investigating the effect of pH, iron concentration and filter media type on iron oxide coating development, iv) comparing the performance of pilot filters operating in floc filtration and adsorptive filtration modes, and v) modelling the adsorptive iron removal in filters to predict the iron breakthrough in filters with new and iron oxide coated sand under anoxic conditions.

An experimental method was developed to measure the adsorption of iron(II) onto filter media and iron hydroxides. It was found that adsorption of iron(II) onto filter media can adequately be described with both the Freundlich and the Langmuir isotherms. The iron(II) adsorption capacities of the different filter media tested varied widely. Of the virgin materials tested, basalt showed the highest iron(II) adsorption capacity followed by anthracite, olivine, magnetite, sand, pumice, and limestone. Iron oxide coated sands from full-scale iron removal plants demonstrated a much higher capacity for iron(II) adsorption compared to new (virgin) sand. In the pH range examined (6-7.5), the iron(II) adsorption capacity of both new and iron oxide coated sand increased with the increase of pH. Among the iron hydroxides tested, lepidocrocite had the highest iron(II) adsorption capacity, followed by amorphous iron hydroxide, ferrihydrite, and goethite. An estimation based on the experimental results indicated that the contribution of iron(II) adsorption onto iron hydroxide flocs on the overall process of adsorptive iron removal, as well as floc filtration iron removal, is probably negligible in iron removal plants.

The high iron(II) adsorption capacity of iron oxide coated sand from iron removal plants can, in principle, be utilised to improve iron removal in filters by switching the governing mode of operation from floc filtration to adsorptive filtration. This can be achieved by bypassing the aeration step and/or reducing the pre-oxidation time to ensure that the majority of the iron enters the filter bed in iron(II) form. In practice, primarily adsorptive iron removal in filters can be realised in two operational modes, namely a) intermittent regeneration mode by operating the filter under anoxic conditions and regenerating the adsorption sites by e.g. backwashing with oxygen-rich water or with a chemical oxidant e.g. $KMnO_4$, and b) continuous regeneration mode by operating the filters under aerobic conditions with limited oxygen concentration in the feed water and/or by limiting the pre-oxidation time.

Within the concentration range examined, NH_4^+, Cl^-, alkalinity, and background ionic strength had no significant effect on iron(II) adsorption onto either new silica sand or iron oxide coated sand. An increase in concentrations of Mn^{2+}, Ca^{2+}, PO_4^{3-}, and silica decreased the iron(II) adsorption onto new sand, whereas an increase in iron(II) adsorption was observed when SO_4^{2-} concentration was increased. The effect of Mn^{2+}, Ca^{2+}, SO_4^{2-} and PO_4^{3-} on iron(II) adsorption onto iron oxide coated sand, however, was minimal. In general, the effect of different parameters on iron(II) adsorption was more pronounced on new sand than on iron oxide coated sand. This indicates that once the iron oxide coating is developed on the surface of the filter media, iron(II) adsorption is not hindered significantly by the presence of other inorganic ions within the concentration range common for groundwater. Preliminary experiments with commercial humic acid showed a negative effect of organic matter on iron(II) adsorption capacity. The effect of organic matter present in groundwater on iron(II) adsorption onto filter media needs further detailed investigation.

Analysis of the physical and surface chemical characteristics of coated sand from twelve groundwater treatment plants in the Netherlands showed that compared to new sand, coated sand had a very high porosity and a very large specific surface area. In general, the iron content of the coating and iron(II) adsorption capacity increased with time in use. However, the average annual increase of the iron content and the adsorption capacity varied for the coated sand from different plants, probably due to the difference in water quality, process conditions applied, and time in use. The grain size of the filter sand increased and the density decreased with the development of iron oxide coating. The decrease in density of coated sand with the iron oxide coating development was a function of the increase in the effective grain size. The measured high adsorption capacities of coated sand from wet filters and dry filters of full-scale groundwater treatment plants indicate that, in wet filters, adsorptive iron removal also plays a role. In dry filters, this mechanism should be dominant due to a very short pre-oxidation time.

The development of an iron oxide coating on the filter media is an important factor in effective adsorptive iron removal from groundwater. The rate of development of the coating and its characteristics are influenced by raw water quality, process conditions applied, and characteristics of the filter media. It was found that preconditioning of new filter media (e.g. at high feed water pH and/or high iron concentration) results in rapid development of an effective iron oxide coating that can reduce the initial filter ripening time. Additionally, the use of virgin media with a high iron(II) adsorption capacity, like basalt, can also reduce the time required to develop a coating with an adequate adsorption capacity.

The process of adsorptive iron removal in filters under anoxic conditions was modelled using adsorption isotherm parameters, mass balance, and mass transfer equations. Experimental results were compared with the predictions of three fixed bed adsorption models, namely i) Constant Pattern Model (CPM), ii) Linear Driving Force Model (LDFM), and iii) Plug Flow Homogeneous Surface Diffusion Model (PFHSDM). The CPM, which considers external mass transfer only, was not sufficient to predict iron breakthrough in filter columns with new and iron oxide coated sand. The LDFM and the PFHSDM predictions of iron breakthrough were more accurate in the case of new sand. In the case of iron oxide coated sand, the predictions were not satisfactory. The difference in model predictions and experimental results in the case of iron oxide coated sand was probably due to the effect of an initial pH drop in the pores of the filter media with iron(II) adsorption, and a consequent decrease in iron(II) adsorption capacity.

Adsorptive filtration can potentially be employed as the primary method of iron removal from anoxic groundwater without manganese and ammonium. This process could also be very attractive in situations where two filtration steps are applied due to high concentrations of iron, manganese, and ammonium in raw water. The first filter can be optimised as an adsorptive iron removal filter, while the second filter can be employed for manganese and ammonium removal.

Adsorptive iron removal is potentially an attractive alternative to conventional floc filtration iron removal. Application of this process has prospects of improving the filtrate quality, extending the filter run time, and easing or reducing the treatment of filter backwash water and sludge, thus resulting in a higher treatment process efficiency.

Key words: groundwater, iron removal, filtration, removal mechanisms, floc formation, adsorption, process efficiency, adsorption capacity, iron oxide coated sand, water quality, coating development, modelling, mass transfer.

Chapter 1

INTRODUCTION

1.1 GROUNDWATER USE AND TREATMENT

1.1.1 Groundwater for water supply

Groundwater has been used as a source of drinking water since time immemorial. In Egypt wells were already being used in 3000 BC (Katko 1997). Excavations at Mohanjodaro have revealed brick-lined dug wells existing as early as 3000 BC during the Indus Valley Civilisation (Raghunath 1987). Wells and their importance can also be traced in the Book of Genesis. The Bible recounts numerous incidents which illustrate the importance of groundwater supplies to the tribes of Israel (Rail 1989).

Groundwater is the major source of drinking water in many countries across the world. Table 1.1 summarises the groundwater use as drinking water in different regions. Groundwater is extensively used as an important source of public water supply in Europe ranging from nearly 100% in Denmark, 72% in Germany, and 56% in France to 27% in the United Kingdom (EEA 1999). In the United States, groundwater is the primary source of potable water for over 96% of the rural population (Biswas 1997). In some Asian countries the share of groundwater in drinking water supplies was as follows: India 80% (rural), Maldives 80%, Philippines 60%, Thailand 50%, and Nepal 60% (Das Gupta 1991). These inventories of groundwater use in water supply reveal its worldwide importance.

Table 1.1 Groundwater as a share of drinking water by region

Region	Share of drinking water From groundwater (%)	People served (millions)
Asia Pacific	32	1000 to 1200
Europe	75	200 to 500
Latin America	29	150
United States	51	135
Australia	15	3
World		1500 to 2000

Source: Sampat (2000)

Groundwater is generally a preferred source for water supplies because of its convenient availability close to where water is required, its constant and good natural quality (which is frequently adequate for potable water supplies with minimal treatment), and relatively low capital cost of water supply system development. Against these common advantages, it should be noted that groundwater is vulnerable to contamination by various anthropogenic activities (agricultural, domestic, and industrial). Contrary to the popular impression that the waters from springs and wells are "pure", patterns of pervasive pollution of groundwaters are being uncovered.

Groundwater is naturally of very good microbiological quality and its chemical quality depends on hydrogeological conditions. Naturally occurring groundwater quality problems are typically associated with high hardness, high salinity and elevated concentrations of iron, manganese, ammonium, fluoride, and occasionally methane, hydrogen sulphide, nitrate, and arsenic (Nash and McGall 1994). Hence, generally some form of treatment will be required for potable water supplies. A high concentration of iron and manganese is by far the most common water quality problem associated with groundwater.

1.1.2 Iron in groundwater

Iron being the fourth most abundant element and second most abundant metal in the earth's crust (Silver 1993; WHO 1996), is a common constituent of groundwater. The presence of iron in groundwater is generally attributed to the dissolution of iron bearing rocks and minerals, chiefly oxides (hematite, magnetite, limonite), sulphides (pyrite), carbonates (siderite) and silicates (pyroxene, amphiboles, biotites and olivines) under anaerobic conditions in the presence of reducing agents like organic matter and hydrogen sulphide (O'Connor 1971; Hem 1989).

Iron usually exists in two oxidation states, reduced soluble divalent ferrous (Fe^{2+} or iron(II)) and oxidised trivalent ferric (Fe^{3+} or iron(III)). Iron may be present in groundwater in the following five forms: i) dissolved as iron(II), ii) inorganic complexes, iii) organic complexes, iv) colloidal, and v) suspended. The state of the iron in water depends above all on the pH and the redox potential (Eh) (Fig. 1.1).

Most natural waters have pH values ranging from 5.0 to 8.5, and pE values ranging from -7 to +12. Thus, iron(II) would be the predominant species in the absence of an electron acceptor such as oxygen (Hem 1989; Faust and Aly 1998). The concentration of iron in natural waters is frequently limited by the solubility of its carbonate. Waters of high alkalinity often, therefore, have a lower iron content than water of low alkalinity (O'Connor 1971; ASCE and AWWA 1990). Iron concentration in groundwater normally ranges from a few hundredths to about 50 mg/l with the majority containing <5 mg/l (Hem 1989; Davis 1997).

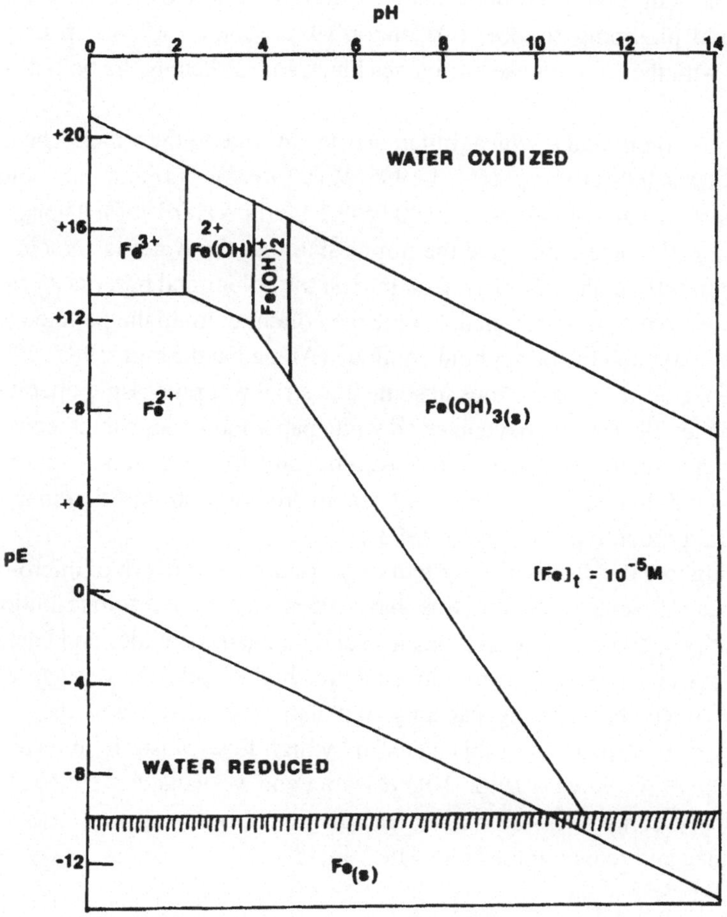

Fig. 1.1 pH-pE diagram for the iron system (Faust and Aly 1998)

The iron problem

There is no health consequence of iron in drinking water. Iron is an essential element in human nutrition. Estimates for the minimum daily requirement for iron depend on age, sex, physiological status and iron bioavailability and range from 10 to 50 mg/day. An intake of 0.4-1 mg iron/kg of body weight per day is unlikely to cause adverse effects in healthy persons. Allocation of 10% of this provisional maximum tolerable daily intake (PMTDI) to drinking water gives a value of about 2 mg/l, which does not present a hazard to health (WHO 1996).

Iron is normally present in groundwater worldwide. Iron in water supplies, however, is undesirable, as it is a nuisance for domestic and industrial users and water suppliers causing various aesthetic and operational problems as listed below.

1. Iron produces ugly and insoluble rusty oxide-red, yellow or brown stains and streaks on laundry and plumbing fixtures (O'Connor 1971; Kothari 1988). In extreme cases, iron interferes with the culinary use turning tea black and darkening the boiled vegetables (Hauer 1950).
2. Iron imparts colour and a typical bitter, astringent taste to the water. The taste threshold of iron in water is 0.04-0.1 mg/l (JMM 1985; WHO 1996). Turbidity and colour may develop in piped systems at iron levels above 0.05-0.1 mg/l (WHO 1996). Though harmless, these organoleptic characteristics give the impression that the water is somehow contaminated. Most importantly in the developing countries, the colour and bitter taste caused by iron can result in well water being rejected. People then often return to the polluted surface water and so incidents of cholera and typhoid continue (Ahmed and Smith 1987, 1988; Chibi 1991).
3. The presence of iron is disastrous in some industrial wet processing operations. Water to be used in the textile, dyeing, beverage and white paper industries should contain less than 0.05 mg/l of iron or manganese (Cox 1964). Additionally, the oxidation of iron-rich water applied to cultivated fields can lead to low-pH ferric hydroxide-rich soils that may severely damage agricultural productivity (Chapelle 1993).
4. Iron passing into the distribution system may promote the growth of micro-organisms. Slime thicknesses of several centimetres have been observed in distribution pipes. These accumulations, consisting of hydrous iron and manganese oxides and bacteria, increase the friction loss and power consumption, require higher chlorine dosage, deplete dissolved oxygen, reduce the carrying capacity and may eventually clog the distribution pipes. Sloughing or resuspension of this material by high flow causes high turbidities. (O'Connor 1971; Culp 1986; Salvato 1992; Vigneswaran and Visvanathan 1995). Therefore, for the water supply companies the main concerns are minimising the costs of operation and maintenance and reducing the "red water" incidents.

Iron removal from groundwater is, therefore, a major concern for most water supply companies using groundwater as their source. To prevent the difficulties mentioned above, various regulatory agencies have put forward standards or guidelines to control iron concentrations in water supplies. An AWWA task group suggested limits of 0.05 mg/l for iron and 0.01 mg/l for manganese for an "ideal quality water" for public use (Bean 1962). Based on taste and nuisance considerations, the World Health Organisation (WHO) recommends that the iron concentration in drinking water should be less than 0.3 mg/l (WHO 1996). The EC directive recommends that the iron in water supplies should be $\leq$ 0.2 mg/l (EC 1998). In the Netherlands, the guideline level for iron in drinking water is $\leq$ 0.05 mg/l (VEWIN 1993) and several water supply companies are aiming at a level of $\leq$0.03 mg/l in order to minimise the distribution system maintenance costs.

1.1.3 Groundwater use and iron removal practice in the Netherlands

The kingdom of the Netherlands is situated along the North Sea in north-west Europe covering a land area of approximately 34,000 square kilometres. The total territory, including inland lakes, estuaries and territorial sea, amounts to 41,160 square kilometers. The topography is relatively flat with the elevation ranging from -6.7 m to 322 m above mean sea level.

The population of the Netherlands is about 15.9 million (2000), with an average population density of 460 per square kilometer, of which 99% have access to public water supply services (CBS 2001).

In the Netherlands, there are at present about 20 water supply companies with a total of 250 treatment facilities and an annual production of 1300 x 10^6 m^3. A total of 232 groundwater treatment plants, of which five abstract a mixture of groundwater and riverbank infiltrate, produce 805 x 10^6 m^3/year (van der Kooij *et al.* 1999). Groundwater contributes to about 62% of the water produced by water companies in the Netherlands (CBS 2001). The removal of iron, manganese, ammonium and methane is the most important step in groundwater treatment. Groundwater extracted in the Netherlands contains 0-30 mg/l Fe (mean 4.8 mg/l), 0-2 mg/l Mn (mean 0.2 mg/l) and 0-35 mg/l NH_4^+ (mean 0.6 mg/l). The removal of these constituents is carried out by a combination of aeration and filtration. Several aeration systems, namely: cascades, spray aeration, tower aeration (co-current and counter current), venturi aeration, and plate aeration are used for iron oxidation. Some plants also use potassium permanganate and ozone as oxidant, and aluminium or iron salts as coagulant to facilitate iron removal. The single or dual media rapid filters used are of open gravity, pressurised, and dry type (van Wijk *et al.* 1987; Kruithof and Koppers 1989). Many groundwater treatment plants employ two filtration steps for iron, manganese and ammonium removal. Iron is removed in the first filter and then manganese and ammonium are removed in the second filter. In the Netherlands, treatment of the filter backwash water is gradually becoming a common practice and a beneficial application of the sludge after thickening is now being introduced.

1.1.4 Groundwater use and iron removal practice in Nepal

Nepal is a landlocked, mountainous developing country in South Asia covering an area of 147,181 square kilometres. It is characterised by diverse physiographic zones, contrasting climates and attitudinal variations ranging from 75 to 8848 m. High mountains and rolling hills account for 83% of the total land area and the remaining 17% are occupied by the plains of Terai (CBS 1998). The estimated population in 2000 is 22.9 million and the growth rate is 2.4% per annum (MOPE 2000). Despite the high priority awarded to water supply and sanitation in Nepal and the rising public investments over the last decade, major shortfalls in the level and quality of service coverage still remain. The estimated national coverage of water supply services by the end of July 2000 is 61% (DWSS 2000).

A survey conducted by ESCAP in 1990 revealed that in Nepal 60% of drinking water, 80% of the municipal water supply and 20% of agricultural water supply comes from underground aquifers (Das Gupta 1991). Groundwater is extensively used for water supply in the Terai plains of Nepal and in the Kathmandu valley. In the Terai region, shallow hand pumps and deep tube wells are used extensively for the provision of drinking water. Iron in groundwater is a major quality concern in many rural areas where hand pumps and deep wells are used. People often complain of the bitter taste and the colour of the water. Some people still go back to the traditional surface water sources to collect drinking water and use water from wells for purposes other than drinking. Hand pumps and tube wells frequently fail to provide the required service due to clogging by iron deposits.

Attempts have been made to solve the problem in some areas by installing low-cost iron removal units (with a perforated pipe for aeration and twin chambers for sedimentation and filtration) promoted by UNICEF (UNICEF 1986). However, they are very few in number and the majority of them do not function properly or are not in use because of various operational problems. Some households practice domestic iron removal using three pitcher-sand filtration. Very often, hand pumps are abandoned or rejected shortly after installation by the government or donor agencies for other water sources with inferior water quality because of the colour and taste associated with groundwater from hand pumps. Thus, in the absence of proper iron removal methods, many water supply systems installed have failed to give the intended health benefits of safe drinking water.

Groundwater provides for about 40% of the water needs of Kathmandu valley towns (population about 1.1 million) during the dry season with the annual average contribution of about 25% (JICA 1990; ADB 1997; BTW 1998). Integrated surface water-groundwater treatment using the extensive treatment processes of bio-filtration, flocculation/sedimentation, and rapid sand filtration has been employed in Mahakalchaur and Bansbari water treatment plants in Kathmandu. In some Terai towns, namely Rajbiraj, Lahan, Damak, Inarwa, Kakarbhitta, Birtamod, Chandragadhi and Bhadrapur, some iron removal plants have been constructed which employ the process of aeration/chlorination followed by rapid filtration. Problems associated with iron are, however, often reported due to operational inconveniences. In many cases, the raw groundwater is often pumped directly into the distribution network. The lack of knowledge of the mechanisms of iron removal and the relatively high cost and complexity of providing the necessary treatment has led to either inadequate or no treatment.

1.2 IRON REMOVAL METHODS

Treating groundwater to remove iron from municipal, agricultural and domestic wells is a multi-million dollar a year business throughout the world (Chapelle 1993). The first iron removal plant was constructed at Charlottenburg, Germany in 1874. The earliest plants employed aeration and filtration, sometimes supplemented by the addition of lime, to treat groundwaters (O'Connor 1971). The same method of treatment predominates today.

Removal of iron from groundwater can be accomplished in several ways. The type of treatment largely depends on the quality of the raw water, financial resources available and the philosophy of the water company. The following methods are used to control iron in the water supply:

1. Oxidation-precipitation-filtration
   (a) Oxidation by aeration, sometimes with a detention or sedimentation tank and

the addition of chemicals for pH adjustment.
(b) Chemical oxidation using chlorine, chlorine dioxide, ozone, potassium permanganate or hydrogen peroxide (with or without pre-aeration).
(c) Biological oxidation

2. Ion exchange or zeolite softening
3. Stabilisation or sequestering using silicates or polyphosphates
4. Lime softening or limestone bed filtration
5. Manganese greensand process
6. In situ oxidation (subsurface iron removal or VYREDOX process)
7. Membrane processes
8. Calcined magnesite - diatomaceous earth filtration (O'Connor and Benson 1970)
9. Sirofloc (activated magnetite) process (Gregory *et al.* 1988; Home *et al.* 1992)
10. Catalytic or adsorptive filtration using patented filter media impregnated with various oxides of iron and/or manganese like BIRM, PYROLOX, Anthrasand, Pyrolusite, Aqua Mandix, Catalytic Carbon, etc. (Sommerfeld 1999)

The suitability, advantages and limitations of some of the most commonly used methods are summarised in Table 1.2

Table 1.2 Iron Removal Methods

Removal Method	Application	Advantage	Limitation
1. Oxidation, precipitation and filtration			
(a) Oxidation by aeration	• Fe <5 mg/l and little or no organic matter or other reducing agents • As pre-oxidation step to save chemical costs when Fe >5 mg/l	• No chemicals required • Simple in operation • Partly removes CO_2, H_2S and CH_4 present	• Ineffective in cases of low pH, and high Fe and Mn concentrations or when Fe is organically complexed • Initial cost is high
(b) Oxidation with chlorine	• Beneficial to remove Fe and Mn in one step • Optimum pH 6.8-8.4 (Kothari 1988)	• More rapid oxidation than aeration especially under conditions of organic matter interference • Less expensive and more effective than $KMnO_4$ • Can also be used for disinfection	• THM formation • Chloro-derivatives can cause taste and odour problems • Require safe handling and storage of chlorine and chlorine compounds

Removal Method	**Application**	**Advantage**	**Limitation**
(c) Oxidation with chlorine dioxide	• Effective when iron is organically complexed or ammonium concentration is high (Twort *et al.* 2000) • Reacts more rapidly with Mn than chlorine	• No THM formation	• Costlier than chlorine • Possible health effects of by-products (Twort *et al.* 2000) • Not used for iron removal only (Culp 1986)
(d) Oxidation with potassium permanganate	• Fe <5 mg/l (Wong 1984) • More efficient at pH >7.5	• Less equipment and capital investment compared to chlorine • Efficient; rapid and complete reaction • Reacts with H_2S, cyanides, phenols, and other taste and odour-producing compounds	• Difficult to control • Overdose (≅ 0.05 mg/l) may produce a pink colour • Ineffective for high iron concentrations • More expensive than chlorine and ozone
(e) Oxidation with Ozone	• Effective even when the iron is organically complexed (Cromley and O'Connor 1976; Paillard *et al.* 1991)	• Powerful and effective oxidant • Multi-purpose applications of ozone e.g. disinfection, color removal, taste and odour control • No THM formation	• High initial capital and operating costs • May oxidise Mn^{2+} to MnO_4^- resulting in a pink colour • Formation of unwanted by-products e.g. BrO_3^-
(f) Oxidation with hydrogen peroxide	• Very effective when iron is organically complexed	• Faster oxidation • Forms dense, easily settled solid • Cheaper than ozone • Leaves no residue	• Formation of AOC
(g) Biological iron removal	• Recommended for groundwater with acidic or neutral pH (Mouchet 1992; Bourgine *et al.* 1994)	• Higher filtration rate • Longer filter run • Reduced capital and O&M costs	• Sensitive to process conditions (pH and temperature dependent) • Ineffective in presence of NH_4^+ and H_2S

Removal Method	Application	Advantage	Limitation
2. Ion exchange (Zeolite softening)	• Suitable for individual water supply scheme with <5 mg/l of Fe (Gass 1977) • Removes dissolved Fe and Mn together with hardness • Used as a polishing step in some plants/ household use after municipal treatment	• Softening occurs with exchange of Ca^{2+} and Mg^{2+}	• Possibility of resin/zeolite fouling or loss of exchange capacity in presence of O_2 due to iron precipitation • High capital cost • Requires skilled personnel • Ineffective for colloidal or complexed iron
3. Stabilisation or Sequestering process (with polyphosphates and silicates)	• Fe should be <1 mg/l (Salvato 1992) • For distribution system corrosion and deposition inhibition • Practical in controlled use only	• Complexes iron and holds it in solution and the consumers do not notice its presence • No sludge generation	• More expensive than Cl_2 and $KMnO_4$ • Phosphate introduced may stimulate biological growth; May require chlorination to prevent bacterial growth • Cold water only (complex break down when heated) • Polyphosphate complexes may degrade after 48-72 hours (Lorenz *et al.* 1988)
4. Lime softening/ Limestone bed filtration	• Pre-aerated water with pH >9.5 and sufficient alkalinity (>20 mg/l as $CaCO_3$).	• Beneficial when a large amount of softened water is required	• Not cost effective unless lime treatment is also required for hardness reduction (Culp 1986) • Increased sludge problems
5. Manganese greensand process	• Removes Fe and Mn by combination of sorption and oxidation • Max Fe + Mn <10 mg/l • H_2S <2-5 mg/l • Optimum pH 6.2-8.0	• H_2S can be removed together with Fe and Mn	• Higher head loss, shorter run time • High O&M costs ($KMnO_4$ regeneration) • Not suitable for larger water treatment plants (JMM 1985)

Removal Method	Application	Advantage	Limitation
6. In-situ oxidation (sub-surface removal/ Vyredox method)	• By infiltrating oxygen-rich water into the ground through a well	• No chemicals • Abstraction-infiltration ratio of groundwater is high	• Potential for contamination of aquifer • Clogging of the aquifer, corrosion of well screen • Excessive bacterial growth may occur around the well.
7. Membrane processes	• NF/RO - to remove dissolved iron • MF/UF - to remove iron flocs	• Can be combined with the removal of other constituents e.g. hardness and THM precursors	• High capital and O&M costs • MF/UF membranes require frequent cleaning

Among the different techniques mentioned above, aeration or chemical oxidation followed by rapid sand filtration is most widely used (O'Connor 1971; Wong 1984; JMM 1985; Culp 1986; Salvato 1992; Twort *et al.* 1994; Sommerfeld 1999). Aeration - rapid sand filtration is the preferred method in the Netherlands and in developing countries. Compared to other methods, this method is more economical, less complicated and generally avoids the use of chemicals, which is not usually welcome in the water industry.

1.3 CHEMISTRY OF IRON REMOVAL

1.3.1 Iron oxygenation kinetics

Iron oxidation and its removal is based on the transformation of the soluble form of iron (Fe^{2+}) to an insoluble form (Fe^{3+}). In simplified notation,

$$4Fe^{2+} + O_2 + 2H_2O \rightarrow 4Fe^{3+} + 4OH^- \qquad (1.1)$$

$$4Fe^{3+} + 4OH^- + 8H_2O \rightarrow 4Fe(OH)_3 + 8H^+ \qquad (1.2)$$

$$4Fe^{2+} + O_2 + 10H_2O \rightarrow 4Fe(OH)_3 + 8H^+ \qquad (1.3)$$

Equation 1.1 shows that about 0.14 mg of oxygen is required for the oxidation of 1 mg of iron(II). Therefore, the oxygen concentration in aerated water is theoretically sufficient for the complete oxidation of iron(II) normally present in natural groundwaters.

It is important to note that iron hydroxides formed after oxidation of iron(II) and subsequent hydrolysis of iron(III) is a complex of different iron hydroxide species and the representation of $Fe(OH)_3$ is merely a simplification of reality. The hydrolysis of iron(III) is discussed in detail in section 1.3.3.

Iron oxygenation kinetics has been extensively studied (Stumm and Lee 1961; Lerk 1965; Jobin and Ghosh 1972; Olson and Twardowski 1975; Sung and Morgan 1980; Robinson *et al.* 1981; Davison and Seed 1983). Stumm and Lee (1961) found that the rate of oxygenation of ferrous iron in bicarbonate solutions is of the first order with respect to both the concentrations of iron(II) and dissolved oxygen and of the second order with respect to the OH^- ion.

$$\frac{-d[Fe(II)]}{dt} = k_o \, pO_2 \, [Fe(II)] \, [OH^-]^2 \qquad (1.4)$$

where d[Fe(II)]/dt = rate of iron(II) oxidation (mol l^{-1} min^{-1})
k_o = reaction rate constant = 8.0 (± 2.5) x 10^{13} l^2 mol^{-2} atm^{-1} min^{-1} at 20.5°C
pO_2 = partial pressure of oxygen (atm) = 0.21 $[O_2]/[O_{2\text{-sat}}]$
$[O_2]$, $[O_{2\text{-sat}}]$ = actual and the saturated concentration of oxygen in water (g/m^3).
[Fe(II)] = concentration of ferrous iron (mol/l).
$[OH^-]$ = concentration of hydroxyl ions (mol/l)

Equation (1.4) shows that the oxygenation rate is very strongly pH-dependent, increasing 100-fold for each unit increase in pH. Therefore, the rate of oxidation of iron(II) is slow at low pH. Their studies also showed that oxidation of ferrous iron should be expected to occur rapidly in well-oxygenated waters at pH values exceeding 7.2.

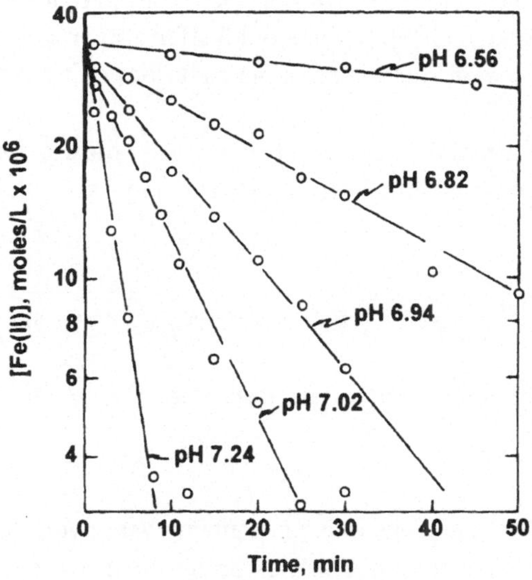

Fig. 1.2 Oxygenation rate of ferrous iron is proportional to Fe(II) and is strongly influenced by pH (Stumm and Lee 1961)

Recent studies have shown that in the presence of iron(III), the oxidation of iron(II) takes place via two parallel paths. One of these is the homogenous reaction occurring in the solution and the other is the heterogeneous reaction occurring on the surface of iron hydroxide precipitates. (Tamura *et al.* 1976; Tufeckci and Sarikaya 1996). At constant pH and O_2 concentration, the rate equation is given by

$$\frac{-d[Fe(II)]}{dt} = (k + k' Fe[III])[Fe(II)] \tag{1.5}$$

where k = rate constant for the homogeneous reaction = $k_o\,[O_2]\,[OH^-]^2$

k' = rate constant for the heterogeneous reaction = $k_{so}\,[O_2]K/[H^+]$

k_o and k_{so} are the real rate constants for the reactions and K is the equilibrium constant for the adsorption of iron(II) on iron(III) hydroxide. The numerical values of the constants are $k_o = 2.3 \times 10^{14}\ l^3\ mol^{-3}\ s^{-1}$, $k_{so} = 73\ l\ mol^{-1}\ s^{-1}$ and $K = 10^{-9.6}\ mol\ l^{-1}\ mg^{-1}$ (Tamura *et al.* 1976).

The reaction is, therefore, autocatalytic as the oxidation of iron(II) is facilitated by the reaction-product iron hydroxides. The effect becomes noticeable at iron(III) concentrations exceeding 5~10 mg/l (Tamura *et al.* 1976; Sarikaya 1980) and the oxidation rate reaches a maximum at iron(III) concentration of about 600 mg/l (Tufekci and Sarikaya 1996).

At near neutral pH, most of the iron(III) is in the form of hydroxide precipitate with a positive surface charge. Consequently OH^- is attracted into the diffuse layer; therefore, the pH of the diffuse layer of iron(III) flocs is higher than that of the bulk solution. Thus iron(II) adsorbed on the surface of iron(III) precipitate is oxidised at much higher rates since it is known that the oxidation rate is proportional to the squares of $[OH^-]$ concentration (Tamura *et al.* 1976; Sarikaya 1980; Tufekci and Sarikaya 1996). This could offer an explanation for higher iron removal efficiency in the presence of iron(III) precipitates or iron oxide coatings on the media.

Barry *et al.* (1994) presented the general rate expression for iron oxidation kinetics considering homogeneous oxygenation, abiotic heterogeneous catalysis, biotic oxidation process and other mechanisms.

$$\begin{aligned}\frac{-d[Fe(II)]}{dt} &= \{k_0\,[Fe^{2+}] + k_1[Fe(OH)^+] + k_2[Fe(OH)_2]\}pO_2 \\ &\quad + k'_3\,A[Fe(II][OH^-]^2\,pO_2 + k_4[Bacteria][Fe(II)][OH^-]^2\,pO_2 + R_{other}\end{aligned} \tag{1.6}$$

where

k_0-k_2 = first order homogeneous rate constants in water, adjusted for the presence of ligands and catalysts active in the homogeneous oxidation process ($s^{-1}\ atm^{-1}$)

k'_3 = overall rate constant for heterogeneous, abiotic processes ($l^2\ mol^{-2}\ m^{-2}\ s^{-1}\ atm^{-1}$)

A = heterogeneous, abiotic surface area (m^2)

k_4 = rate constant for some bacteria, the term repeated for various strains and corrected for specific conditions (units are $l^3\ mol^{-2}\ s^{-1}\ atm^{-1}\ cells^{-1}$ for bacteria measured in cells/l or $l^3\ mol^{-2}\ s^{-1}\ atm^{-1}\ g^{-1}$ for bacteria measured as grams volatile solids/l)

[Bacteria] = concentration of bacteria (cells/l or grams volatile solids/l)

R_{other} = the oxidation rate attributable to processes not considered explicitly, such as the presence of reactive species like hydrogen peroxide (mol $l^{-1}\ s^{-1}$)

This suggests that homogeneous oxygenation of iron may often be overshadowed by heterogeneous, biotic and photochemical mechanisms.

1.3.2 Factors affecting oxidation of iron

Besides pH, other water quality parameters like alkalinity (bicarbonate concentration), temperature, organic matter and some elements/ions have also been reported as having a significant effect on the rate of oxidation of iron(II).

Alkalinity

Alkalinity is important in iron removal as it provides the buffer capacity to avoid excess pH drop on iron oxidation and influences the characteristics of the precipitate formed. Iron oxidation and removal is poor at low alkalinity due to slow oxidation and poor floc formation (Robinson and Breland 1968; Hult 1973). Ghosh *et al.* (1966) reported that in groundwaters with high alkalinity (>250 mg/l as $CaCO_3$) the precipitates formed after aeration are primarily carbonates rather than hydroxides, and a large part of the iron precipitated is in the ferrous rather than the ferric form. Cleasby (1975) indicated that the more rapidly the iron is oxidised, that is, through the use of strong oxidants such as permanganate, chlorine or ozone, the more likely it is that the end product will be hydroxide. However, when the oxidation proceeds more slowly with aeration, then most likely the end product will be carbonate in water of high alkalinity.

Jobin and Ghosh (1972) found that buffer intensity (β in eq/pH) of water influences the rate of iron oxidation at values higher than 4.0 x 10^{-3} eq/pH and suggested the following rate equation:

$$\frac{-d[Fe(II)]}{dt} = k\ pO_2\ [Fe(II)]\ [OH^-]^2\ [\beta]^n \qquad (1.7)$$

where $\beta = 2.3\ \{[H^+] + [OH^-] + C_T\ [\alpha_1(\alpha_0 + \alpha_2) + 4\alpha_2\alpha_0]\}$

$C_T = [H_2CO_3] + [HCO_3^-] + [CO_3^{2-}]$

$$\alpha_0 = \frac{[H_2CO_3]}{C_T}, \qquad \alpha_1 = \frac{[HCO_3^-]}{C_T} \quad \text{and} \quad \alpha_2 = \frac{[CO_3^{2-}]}{C_T}$$

(All concentrations are expressed in moles per litre).

Therefore, β depends on pH, temperature, alkalinity, and dissolved solids.

Temperature
Stumm and Lee (1961) observed that at constant pH and oxygen concentration, the rate of oxygenation increases tenfold for a 15°C increase in temperature. However, when Sung and Morgan (1980) normalised the experimental data with respect to changes in K_w and O_2 solubility, the rate constant varied only slightly with increasing temperature.

Organic matter
Iron can be complexed by humic and fulvic acids and similar organic substances present in water. Such complexation may render the iron resistant to oxidation even in the presence of dissolved oxygen (Oldham and Gloyna 1969; Jobin and Ghosh 1972; Knocke *et al.* 1992, 1994). Theis and Singer (1973, 1974) showed that iron(II) complexation by humic matter increases with the increase of organic matter concentration and with the increase in pH. Furthermore, their study also showed that humic substances are capable of reducing iron(III), which depends on both the pH and the relative concentration of humic substances to iron(III).

Catalytic effect of some elements
Different ions present in water can alter the rate of homogeneous oxygenation of iron(II). Accelerating effects have been observed for Cu^{2+}, Mn^{2+}, Co^{2+} and $H_2PO_4^-$ (Stumm and Lee 1961) while inhibiting effects have been reported for SO_4^{2-} and Cl^- (Sung and Morgan 1980). Barry *et al.* (1994), in their work on iron oxidation kinetics in aquatic ecosystems, also found that TiO_2 accelerates the oxidation of iron(II) by forming complexes with it.

Silica in groundwater could interfere with the hydrolysis of oxidised iron and thus hinder filtration (Robinson 1975). Schenk and Weber (1968) reported that dissolved silica (H_4SiO_4 or $Si(OH)_4$) affects the chemical behaviour of Fe^{2+} by catalyzing the rate of iron(II) oxidation. Dart and Foley (1970) present some operational experiences that appear to be opposite to the conclusions drawn by Schenk and Weber (1968). Iron removal problems were experienced in waters with 30 or 40 mg/l silica that often released very little of their iron content on either aeration or chlorination followed by filtration. The silica apparently reacts with $Fe(OH)_3$ and holds it in suspension (See Table 1.2-3. Stabilisation).

1.3.3 Hydrolysis of iron(III)

Iron(III) formed on oxidation of iron(II) subsequently undergoes hydrolysis resulting in the formation of hydrated iron oxide ($Fe_2O_3.xH_2O$). The aqueous chemistry of iron is rather complex since this metal enters into several protolysis and oxidation-reduction reactions (Hem and Cropper 1962; Faust and Aly 1998; Stumm and Morgan 1996). Singley *et al.* (1967, 1969) demonstrated the existence of an entire family of iron hydrates having up to six molecules of water associated with one Fe^{3+} ion.

$$Fe^{3+} + mH_2O \equiv Fe(OH)_m^{(3-m)} + mH^+ \quad (1.8)$$

Lerk (1965) suggested that the oxidation and hydrolysis reactions of iron(II) occur as follows:

$$4Fe^{2+} + O_2 + 2H_2O \rightarrow \quad 4Fe^{3+} + 4OH^- \qquad (1.9)$$

$$4Fe^{3+} + 4OH^- + 2(x+1)H_2O \rightarrow \quad 2(Fe_2O_3 xH_2O) + 8H^+ \qquad (1.10)$$

or at high pH:

$$4Fe^{2+} + 8OH^- \rightarrow \quad 4Fe(OH)_2 \qquad (1.11)$$

$$4Fe(OH)_2 + O_2 + (2x-4)H_2O \rightarrow \quad 2(Fe_2O_3 xH_2O) + 8H^+ \qquad (1.12)$$

The overall reaction can be written as

$$4Fe^{2+} + O_2 + (2x+4)H_2O \rightarrow \quad 2(Fe_2O_3 xH_2O) + 8H^+ \qquad (1.13)$$

The hydrolysed species of iron ions will condense to form dimers through hydroxo- and oxo-bridging. These are called "olation" and "oxolation" respectively. The dimers may undergo additional hydrolytic reactions that could provide additional hydroxo groups, which then could form more bridges. These processes lead to the formation of polynuclear hydroxy complexes and ultimately to the formation of precipitates.

Olation (hydroxo-bridging)

$$2[Fe(H_2O)_5OH^{2+}] \rightarrow \underbrace{[(H_2O)_4Fe - (OH)_2 - Fe(H_2O)_4]^{4+}}_{\text{Dimer}} + 2H_2O \qquad (1.14)$$

Oxolation (oxo-bridging)

$$2[Fe(H_2O)_5OH^{2+}] \rightarrow \underbrace{[(H_2O)_5Fe - O - Fe(H_2O)_5]^{4+}}_{\text{Dimer}} + H_2O \qquad (1.15)$$

1.3.4 Chemical oxidation

The atmospheric oxygen, which is introduced into water during aeration, is usually effective in the oxidation of iron(II). However, when the iron is organically complexed, aeration alone is not sufficient. Secondly, iron oxidation is very slow at pH <7.0. Alternative oxidants like potassium permanganate, chlorine or chlorine dioxide, ozone and hydrogen peroxide could be employed for iron oxidation when aeration is not satisfactory. Chemical oxidation is frequently applied when iron and manganese are to be removed simultaneously in a single filtration step as manganese oxygenation is very slow at pH <9.5.

The oxidation of iron(II) by different oxidants can be described by the following chemical reactions:

$$3Fe^{2+} + KMnO_4 + 7H_2O \rightarrow 3Fe(OH)_3 + MnO_2 + K^+ + 5H^+ \qquad (1.16)$$

$$2Fe^{2+} + Cl_2 + 6H_2O \rightarrow 2Fe(OH)_3 + 2Cl^- + 6H^+ \qquad (1.17)$$

$$Fe^{2+} + ClO_2 + 3H_2O \rightarrow Fe(OH)_3 + ClO_2^- + 3H^+ \qquad (1.18)$$

$$2Fe^{2+} + H_2O_2 + 4H_2O \rightarrow 2Fe(OH)_3 + 4H^+ \quad (1.19)$$

$$2Fe^{2+} + O_3 + 5H_2O \rightarrow 2Fe(OH)_3 + O_2 + 4H^+ \quad (1.20)$$

In each of the above cases, hydrogen ions are produced. Thus, based on these reactions alone, pH may decrease significantly in the absence of sufficient buffer capacity.

Very little has been reported in literature concerning the kinetics of oxidation of iron(II) with chemical oxidants (Willey and Jennings 1963; Abukhudair 1989; Knocke *et al.* 1991). As the chemical oxidation of iron is quite rapid at a pH of 7 or higher, the kinetic considerations have little influence on either facility design or operation (Benefield and Morgan 1990).

1.4 IRON REMOVAL MECHANISMS IN FILTERS

Different mechanisms (physical, chemical, and biological) may contribute to iron removal in filters but the dominant one depends on the physical and chemical characteristics of the water and process conditions (Lerk 1965; Rott 1985; Hatva 1988,1989; Mouchet 1992; Søgaard *et al.* 2000).

1.4.1 Oxidation-floc formation

Oxidation-floc formation (floc filtration) is the conventional approach for iron removal from groundwater. In this method soluble iron(II) present in anoxic groundwater is oxidised to insoluble iron(III) and after precipitation, iron hydroxide flocs are removed in the filters. The removal process consists of the following steps (Rott 1973):

1. Oxidation of Fe^{2+} to Fe^{3+} by aeration or by a chemical oxidant
2. Hydrolysis of Fe^{3+} to iron hydroxides
3. Flocculation/agglomeration of the hydroxide particles.
4. Removal of flocs in rapid sand filters.

This process is pH-dependent and dominant at pH values above 8.5. Under this condition, the oxidation is rapid and flocs are formed prior to entering the filter bed.

Various problems have been encountered in the application of this mechanism. In some plants complete oxidation is not achieved, whereas in others filterable flocs (precipitates) could not be formed (O'Connor 1971; Mouchet 1992). Dissolved iron(II) remaining and colloidal iron(III) formed can both pass the filter, consequently lowering the efficiency of iron removal. The rapid head loss development due to clogging of filters and rapid deterioration of filtrate quality are often responsible for short filter runs and frequent backwashing cycles of iron removal plants. Additionally, filter ripening after backwashing takes a rather long time and a large volume of sludge is produced which must be treated and/or disposed of.

The problem is severe when iron is organically complexed as aeration alone can not oxidise the complexed iron. To overcome the problems associated with oxidation of iron by aeration, strong oxidising agents such as chlorine, chlorine dioxide, potassium permanganate and ozone can be used. These chemicals also need sufficient detention time to allow the oxidation reaction to complete and, in addition, some of them can form unwanted by-products (Culp 1986). Provision of sufficient detention time also requires increased construction costs of the plant. Moreover, if the groundwater pH is low, chemicals are required to raise the pH and to enhance the oxidation, again associated with increased operation and maintenance costs.

1.4.2 Adsorption-oxidation

In the adsorption-oxidation (adsorptive filtration) mechanism, the iron(II) present in anoxic groundwater is removed by adsorption onto the surface of the filter media. Subsequently, in the presence of oxygen, the adsorbed iron(II) is oxidised forming a new surface for adsorption. In this way the process continues. The method therefore relies on the iron(II) adsorption capacity of the filter media. In conventional filters, the iron entering the filter bed in iron(II) form is removed through the adsorption-oxidation mechanism. Iron(II) can also adsorb on iron hydroxide flocs commonly present in the filter. Adsorption-oxidation is also the dominant iron removal mechanism in dry filters and sub-surface iron removal (van Beek 1983; Rott 1985; Braster and Martinell 1988; Appelo *et al.* 1999). For the adsorption mechanism to dominate, pre-oxidation of iron(II) before filtration must be minimal. This can be achieved by reducing the oxidant concentration or the time available for the oxidation reaction. It should also be noted that adsorptive iron removal is only feasible for the removal of iron(II).

To achieve principally adsorptive iron removal, the filters can be operated in the following two modes:

(a) in *intermittent regeneration mode*, filters are operated under anoxic conditions. Oxidation of iron(II) is consequently suppressed by avoiding aeration. After the exhaustion of the iron(II) adsorption capacity of the filter media, the anoxic bed requires regeneration of the adsorption sites by oxidation of adsorbed iron(II). This can be achieved by backwashing the filter with oxygen-rich water or with a chemical oxidant e.g. $KMnO_4$;

(b) in *continuous regeneration mode*, filters are operated under aerobic conditions to allow continuous regeneration of the exhausted adsorption sites. A low concentration of oxygen and/or a short pre-oxidation time is required to avoid the formation of iron hydroxide flocs. In this mode there are three possible options: a) dry filter, b) normal rapid filter operated at a high filtration rate and low depth of supernatant, and c) normal rapid filter with low oxygen concentration (1-2 mg/l) in the feed water. Some iron flocs, however, will also be formed under these conditions and therefore backwashing will be required when maximum head loss is reached.

Many researchers studied the functioning of iron removal plants and observed that the iron oxide coating often plays an important role in the oxidation and removal of iron (Hauer 1950; Cox 1964; O'Connor 1971; Anderson *et al.* 1973). This was evident from the improved iron removal after a coating of iron oxide had developed on the filter media. Some researchers have termed this as "**Catalytic Iron Removal**" because the iron removal rate increases with the formation of coating as the previously retained iron oxide acts as the "catalyst" for further iron removal. Cox (1964) found that filters may serve as contact beds following aeration where "catalytic action" of previously precipitated iron oxides will facilitate the oxidation of iron. From their study of iron removal in filters, Ghosh *et al.* (1967) concluded that a fraction of the ferrous iron might have been adsorbed onto the ferric hydrate precipitates. O'Connor (1971) reported that precipitates of hydrous oxides of iron(III) formed after oxidation and deposited on the filter sand serve as adsorption media for iron still in solution. The iron hydroxides have high adsorption capacities for iron(II), thus accounting for the improved removal when filters are ripened and deposition of the precipitates have taken place.

Previous researches at IHE (Adekoya 1995; Amoateng 1996; Ibrahim 1997) showed that water quality improves, filter ripening time reduces, and filter run time increases when iron(II) ions are removed via adsorption onto filter media, compared when iron(III) are removed as flocs. Ibrahim (1997) reported that iron oxide coated sand has much higher efficiency for iron removal compared to new sand. Further, it was found that once the coating is developed, the filter media could give a higher efficiency even at lower pH. Sharma (1997) found that compared to new filter sand, iron oxide coated sand has very high capacity for iron(II) adsorption. This indicates the possibility of improving efficiency of iron removal in the filter by maximising the adsorption of iron(II) onto iron oxide coated media. Adsorption of iron(II) onto iron oxide coated media could be the primary iron removal method for treating anoxic groundwater and an attractive alternative to the conventional oxidation-floc formation method. With the adsorption-oxidation mechanism, the head loss is likely to be very low because the iron forms a coating on the filter media, rather than a floc which blocks the filter pore. Thus, the filter runs could be longer and the backwash water requirement and volume of the sludge reduced. Under this mechanism, it is likely that filters could be run at higher filtration rates as head loss development is not a limitation. Hence, considerable savings are likely in the capital, and operation and maintenance (O&M) costs.

1.4.3 Biological iron removal

Biologically mediated oxidation and removal of iron has been reported in rapid sand filtration of groundwater (Frischherz *et al.* 1985; Czekalla *et al.* 1985; Badjo and Mouchet 1989; Hatva 1989; Mouchet 1992; Bourgine *et al.* 1994). Biological iron removal mainly depends on the activities of microorganisms, which have the unique property of causing oxidation and precipitation of dissolved iron under pH, and redox potential (Eh) conditions that are intermediate between those of natural groundwater and those required for conventional (physical-chemical) iron removal. Figure 1.3 shows the field of activity of these bacteria which straddles the theoretical boundary between the fields of Fe^{2+} stability and the formation of iron hydroxides as defined by thermodynamic analysis of the electrochemical equilibria.

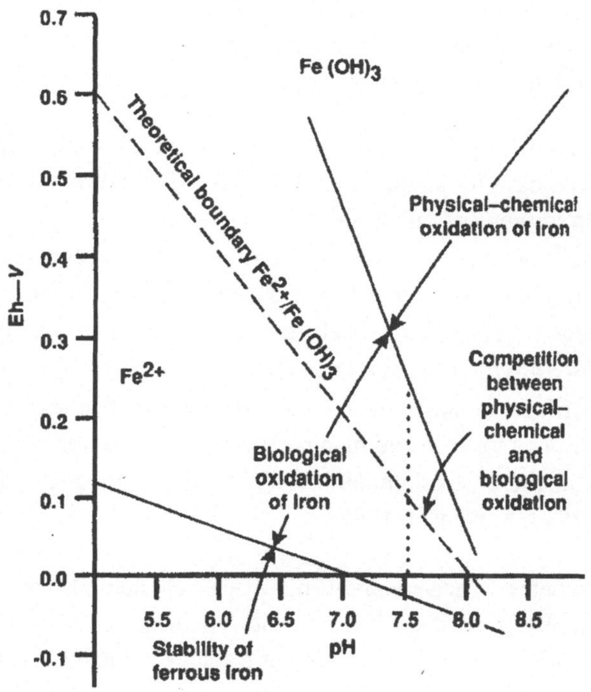

Fig. 1.3 Field of activity of iron bacteria (Mouchet 1992)

The exothermic oxidation of iron(II) can be catalysed by some bacteria due to the oxidation-reduction enzymes which they excrete (flavins); trivalent iron rendered insoluble in hydroxide form is then stored in the mucilaginuos secretions (sheaths, stalks, capsules etc.) of these bacteria. The organisms responsible for this phenomenon are *Gallionella, Leptothrix, Crenothrix, Clonothrix, Siderocpasa, Sphaerotilus, Ferrobacillus and Sideromonas* (Degremont 1991). These iron-oxidising bacteria are widespread and are prevalent in groundwater, ponds, hypolimnion of lakes or impoundments, sedimentary deposits and soil. Two mechanisms of bacterial oxidation have been reported (Czekalla *et al.* 1985; Bourgine *et al.* 1994):

i) Intracellular oxidation by enzymatic action (*Gallionella and Leptothrix ochracea*),
ii) Extracellular oxidation by the catalytic action of excreted polymers (*Gallionella, Leptothrix, Crenothrix, Clonothrix, Sphaerotilus, and Siderocapsa)*

A pH of 6-8 is required for their activity. However, at a pH above 7.2, biological processes will compete with conventional (physical-chemical) processes. The optimum temperature typically ranges from 10 to 15°C for *Gallionella ferruginea* and 20 to 25°C for the *Sphaerotilus-Leptothrix* group (Mouchet 1992).

Mouchet (1992) reported the marked improvement in performance by converting conventional iron treatment plants to biological ones. The primary advantages associated with this process are high filtration rates (10-70 m/h), high retention capacity (1-5 kg Fe/m^2), elimination of chemical reagents, flexibility of operation, and reduced capital and operating costs. However, high iron concentrations in the influent may cause breakthroughs, as the rate of absorption by the bacteria may not be high enough to match the supply rate.

The main disadvantage of this process is the long maturation time before full efficiency is achieved; perhaps 50-60 days for a new filter and 5 days after a 2-month shut down (Stevenson 1997). The other shortcomings of this mechanism include:

- anaerobic conditions may develop in the filter bed, thus converting back iron(III) to iron(II) resulting in an elevated iron concentration in the filtrate;
- increased sludge production and backwash water with filter ageing;
- not suitable for all types of groundwater;
- need for two filtration stages to remove iron and manganese as the required redox potential conditions for iron and manganese oxidising bacteria are very different;
- ineffective in the presence of ammonia (NH_4^+) and inhibiting substances like H_2S and Zn (Twort *et al.* 1994; Stevenson 1997).

It is still debated whether "near neutral pH bacteria" actually oxidise iron and grow autotrophically or merely deposit iron in an oxidised form. In iron bacteria other than the acidophile species, autotrophy using ferrous iron as a source of energy has not been conclusively demonstrated (Hughes and Poole 1989). The bacteria do however process iron intracellularly. Environments high in iron(II) will lead to elevated levels inside the cell. The cells have mechanisms to remove such unwanted ions. Therefore, oxidation of iron may be purely to detoxify their intracellular environment rather than to create energy. The cell membrane and extracellular polymers do provide many sites for the adsorption of iron(II) ions. Once adsorbed the ions may undergo the oxidation reaction. This leads to the formation of the characteristic sheaths often cited as evidence for biological iron removal. It appears that bacteria can act as a pathway for iron oxidation either via adsorption onto the cell membrane or via oxidation to detoxify the intracellular environment. It cannot be established if such mechanisms provide a significant removal capacity (Hughes and Poole 1989).

In reality the boundary between physicochemical and biological iron removal is not well defined (Degremont 1991). Iron removal from anoxic groundwater by adsorption onto new or iron oxide coated filter media is possible in the pH-Eh range of the iron bacteria, which indicates that the process is not necessarily biological. In many iron removal plants and pilot studies nearly complete removal of iron has been seen to occur immediately after the beginning of the test. The fact that iron was removed at the very beginning of the filtration process strongly suggests that this process is predominantly a physicochemical reaction. It has been reported that in some plants the removal efficiency has increased after the development of bio films or "coatings" on the surface of the filter media (Mouchet 1992; Bourgine *et al.* 1994). This may be due to the increased adsorption of iron(II) onto coated media, the additional iron(II) oxidation by iron bacteria or both. Indeed, controversy surrounds the current debate as to whether the mechanism is physicochemical or biological in nature (Mouchet 1992). It is our hypothesis that iron removal at pH-Eh range where oxidation-floc formation is not feasible is predominantly adsorption-oxidation and can be explained by existing physicochemical models.

1.5 THE NEED FOR THE RESEARCH

Meeting the increasingly stringent requirements of iron in the drinking water supply in developed countries (e.g. desired level of ≤0.03 mg/l in the Netherlands), and reducing the operation and maintenance costs of distribution systems worldwide will require a more efficient removal or minimisation of the iron currently passing the filters. To improve the efficiency of iron removal, it is necessary that all the mechanisms involved are well understood and optimally utilised. Under the commonly applied treatment conditions in iron removal plants, the oxidation-floc formation mechanism is expected to be dominant. The adsorption-oxidation mechanism (adsorptive iron removal), however, has several potential advantages over the oxidation-floc formation mechanism, namely longer filter run, shorter filter ripening time, less backwash water use, and less sludge production.

If the adsorbent (solid surface) is chosen carefully and the solution chemistry is adjusted appropriately, adsorption-based processes are generally capable of removing metals over a wider pH range and to much lower levels than processes based on precipitation. In addition to offering a more reliable and more efficient removal of uncomplexed metal ions, adsorption processes can often remove inorganically and organically complexed metals that would not be removed by conventional treatment methodology (Benjamin 1983). It is well known that iron oxides can adsorb both inorganic ions (Dzombak and Morel 1990) and humic substances in water (Tipping 1981; Gu *et al.* 1994, 1995). There is growing evidence that iron oxyhydroxides or iron-bearing materials are capable of removing many metals over a wide pH range and with high efficiency (Benjamin 1983; Stenkamp and Benjamin 1994; Benjamin *et al.* 1996).

Iron oxide coated sand has been successfully employed in fixed beds to treat metal-bearing waters and wastewaters (Edwards and Benjamin 1989; Bailey *et al.* 1992; Lai *et al.* 1994; Joshi and Chaudhuri 1996; Benjamin *et al.* 1996). The exploitation of the adsorption capability of hydrous iron oxide coatings on the surface of the sand grains in the filters to remove iron(II) from groundwater may require some modification of the process scheme and process conditions (filter media type and size, bed depth, detention time etc.). No information is, however, available on the effectiveness of the adsorption-oxidation mechanism when used as the dominant mechanism of iron removal from groundwater. Also, little is known about the effect of water quality parameters on iron(II) adsorption, the rate of oxidation of adsorbed iron(II), the process of development of iron oxide coating on filter media, and the characteristics of the coating developed under different water quality and process conditions. In order to get more insight into adsorptive iron removal and to further optimise the process of iron removal in filters, a detailed study of the adsorption-oxidation mechanism was required. This study was mainly focussed on adsorption aspects, however the factors governing the rate of oxidation of adsorbed iron(II) was not investigated. The role of biological processes e.g. biological iron oxidation and nitrification are equally important in the iron removal, however these aspects were not studied in this thesis.

The goal of this research is to examine the potential of adsorptive iron removal as an alternative to the conventional floc formation method and to investigate the factors governing the adsorptive iron removal process, particularly the mechanisms involved under anoxic conditions.

1.6 RESEARCH OBJECTIVES

The objectives of this research are:

1. To measure the iron(II) adsorption capacities of different new and iron oxide coated filter media and different iron (hydr)oxides, and to analyse the effect of pH on iron(II) adsorption capacity.
2. To analyse the effect of major groundwater quality parameters namely Mn^{2+}, Ca^{2+}, NH_4^+, Cl^-, SO_4^{2-}, PO_4^{3-}, silica, alkalinity and background ionic strength on iron(II) adsorption onto new and iron oxide coated sand.
3. To analyse the physical and surface chemical characteristics of new sand and iron oxide coated sand developed under different water quality and process conditions.
4. To study the factors affecting the development of iron oxide coating on the filter media.
5. To compare the performance of floc filtration and adsorptive filtration modes of iron removal by conducting pilot plant experiments.
6. To model the process of adsorptive iron removal in filters and to predict the iron(II) breakthrough in filtration columns with new and iron oxide coated sand at different pH under anoxic conditions.

1.7 OUTLINE OF THE THESIS

The thesis is organised in eight chapters. Chapter 1 introduces the subject and presents a general literature review on the methods of iron removal from groundwater, the chemistry and mechanisms involved, and discusses the importance/need of the study.

Chapter 2 analyses the iron(II) adsorption capacities of different new (virgin) and iron oxide coated media. Further, it discusses the effect of pH on iron(II) adsorption. Additionally, iron(II) adsorption capacities of different iron (hydr)oxides prepared in the laboratory together with the effect of pH and ageing on iron(II) adsorption are also described.

Chapter 3 presents the effect of various groundwater quality parameters on iron(II) adsorption onto new and iron oxide coated sand.
In Chapter 4 the physical and chemical characteristics of iron oxide coated sand obtained from twelve different groundwater treatment plants in the Netherlands are examined. Furthermore, the

increase in grain size and change in density of filter media with iron oxide coating development and their implications for practice are discussed.

The effect of feed water pH, iron concentration and filter media type on the development of iron oxide coating on filter media and their influence on iron(II) adsorption capacity are presented in Chapter 5.

In Chapter 6 the results of the pilot plant experiment are presented, in which the performance of the filters operating under floc filtration and adsorptive filtration modes are compared with that of a full-scale conventional iron removal plant.

Mathematical modelling of the intermittent regeneration mode of adsorptive iron removal in filters and the predictions of the iron(II) breakthrough in filter columns with new and iron oxide coated sand, using different fixed bed adsorption models, are presented in Chapter 7.

Finally, Chapter 8 summarises the results of this study presented in the previous chapters and gives recommendations for practice and for further research.

REFERENCES

Abukhudair, M.Y., Farooq, S. and Hussain, M.S. (1989) Kinetics of ozonation of iron(II) and manganese(II) in a pure water system. *Journal of Environmental Science and Health*, **24** (4), 389-407.

ADB (1997) *Second Water Utilities Data Book for the Asian and Pacific Region.* Asian Development Bank, Manila, Philippines.

Adekoya, O.A. (1995) *Optimisation of Iron Removal in an Aeration/Filtration Process.* M.Sc. Thesis EE 179, IHE, Delft, The Netherlands.

Ahmed, F. and Smith, P.G. (1987) Design and performance of a community type iron removal plant for hand pump tube wells. *Journal Inst. Wat. Engg. Sci.*, **41**, 167-172.

Ahmed, F. and Smith, P.G. (1988) The effect of upgrading a tube well water supply in rural areas of Bangladesh. *Aqua*, (2), 78-80.

Amoateng, P. (1996) *Improvement of Iron Removal in Groundwater Treatment at WMO.* M.Sc. Thesis EE 225, IHE, Delft, The Netherlands.

Anderson, D.R., Row, D.D. and Sindelar, G.E. (1973) Iron and manganese studies of Nebraska water supplies. *Journal AWWA*, **65**, 637-641.

Appelo, C.A.J., Drijver, B., Hekkenberg, R. and de Jonge, M. (1999) Modelling in situ iron removal from ground water. *Ground Water*, **37** (6), 811- 817.

ASCE and AWWA (1990) Iron and manganese removal. In: *Water Treatment Plant Design.* Chapter 11, pp 283-320, Second Edition, McGraw Hill Publishing Company.

Badjo, Y. and Mouchet, P. (1989) Appropriate technologies - example of a large biological iron removal plant in Togo. *Aqua*, **38** (3), 197-206.

Bailey, R.P., Bennett, T. and Benjamin, M.M. (1992) Sorption onto and recovery of Cr(VI) using iron oxide coated sand. *Water Science and Technology*, **26** (5-6), 1239-1244.

Barry, R.C., Schnoor, J.L., Sulzberger, B., Sigg, L. and Stumm, W. (1994) Iron oxidation kinetics in an acidic alpine lake. *Water Research*, **28** (2), 323-333.

Bean, E.L. (1962) Progress report on water quality criteria. *Journal AWWA*, **54** (11), 1313-1331.

Benefield, L.D. and Morgan, J.M. (1990) Chemical Precipitation. In: *Water Quality and Treatment - A Handbook of Community Water Supplies.* Chapter 10, pp 641-708, Fourth Edition, McGraw-Hill Inc.

Benjamin, M.M. (1983) Adsorption and surface precipitation of metals on amorphous iron oxyhydroxides. *Environmental Science and Technology*, **17** (11), 686-691.

Benjamin, M.M., Sletten, R.S., Bailey, R.P. and Bennett, T. (1996) Sorption and filtration of metals using iron oxide coated sand. *Water Research*, **30** (11), 2609-2620.

Biswas, A.K. edited (1997) *Water Resources: Environmental Planning Management and Development.* McGraw Hill.

Bourgine, F.P *et al.* (1994) Biological processes at Saints Hill water treatment plant, Kent. *Journal of IWEM*, 8 August 1994, 379-392.

Braester, C. and Martinell, R. (1988) The Vyredox and Nitredox methods of in situ treatment of groundwater. *Water Science and Technology*, **20** (3), 149-163.

BTW (1998) *Preparation of a Management Contract for the Urban Water Supply and Sanitation Services in the Kathmandu Valley: Water Demand Report.* Binnie Thames Water - Management & Operation Consultants.

CBS (1998) *Statistical Pocket Book 1998.* Central Bureau of Statistics, Nepal.

CBS (2001) *Statistisch Jaarboek 2001.* Centraal Bureau voor de Statistiek, The Netherlands.

Chapelle, F.H. (1993) *Groundwater Microbiology and Geochemistry.* John Wiley & Sons Inc. USA.

Chibi, C. (1991) Design and performance of a community level iron removal plant. *Waterlines*, **10** (2), 9-10.

Cleasby, J.L. (1975) Iron and Manganese Removal - A case study. *Journal AWWA*, **67** (3), 147-149.

Cox, C.R. (1964) *Operation and Control of Water Treatment Processes*. WHO, p 209- 218.

Cromley, J.T. and O'Connor J.T. (1976) Effect of ozonation on the removal of iron from groundwater. *Journal AWWA*, **68** (6), 315- 319.

Culp;Westner;Culp (1986) *Handbook of Public Water Systems*. Van Nostrand Reinhold Company, New York. Chapter 19, p 633-644.

Czekalla, C., Mevius, W. and Hanert, H. (1985) Quantitative removal of iron and manganese by microorganisms in rapid sand filters (in situ investigations). *Water Supply*, **3** (1), 111-123.

Dart, F. J. and Foley, P. D. (1970) Preventing iron deposition with sodium silicate. *Journal AWWA*, **62** (10), 663-668.

Das Gupta, A. (1991) Study on groundwater quality and monitoring in Asia and the Pacific. In: *Groundwater Quality and Monitoring in Asia and the Pacific*. Water Resources Series No 70, ESCAP, Bangkok, p 13-99.

Davis, J. (1997) Removing iron and manganese from natural waters. *Plant Engineering*, October, 80-84.

Davison, W. and Seed, G. (1983) The kinetics of the oxidation of ferrous iron in synthetic and natural waters. *Geochimica et Cosmochimica Acta*, **47** (1), 67-69.

Degremont (1991) *Water Treatment Handbook*. Vol. 1 and 2, Sixth Edition, Degremont, France.

DWSS (2000) *Annual Progress Report 1999/00*. Department of Water Supply and Sewerage, HMG of Nepal.

Dzombak, D.A. and Morel, F.M.M. (1990) *Surface Complexation Modelling - Hydrous Ferric Oxides*. John Wiley and Sons, Inc.

EC (1998) *Official Journal of the European Communities*. December 5 1998; Council Directive 98/83/EC on the quality of water intended for human consumption. L330/32 - L330/50.

EEA (1999) *Groundwater quality and quantity in Europe*. June 1999, European Environmental Agency, Copenhagen, Denmark.

Edwards, M. and Benjamin, M.M. (1989) Adsorptive filtration using coated sand a new approach for the treatment of metal bearing wastes. *Journal WPCF*, **61** (9), 1523-1533.

Faust, S.D. and Aly, O.M. (1998) *Chemistry of Water Treatment*. Second Edition, Ann Arbor Press Inc., USA.

Frischherz, H., Zibuschka, F., Jung, H. and Zerobin, W. (1985) Biological elimination of iron and manganese. *Water Supply*, **3** (1), 125-136.

Gass (1977) Removal of iron from a water supply. *Water Well Journal*, **31** (10), 35-36.

Ghosh, M.M., O'Connor, J.T. and Engelbrecht, R.S. (1966) Precipitation of iron in aerated groundwater. *Journal of Sanitary Engineering Division- ASCE*, **90** (SA1), 199-213.

Ghosh, M.M., O'Connor, J.T. and Engelbrecht, R.S. (1967) Removal of iron from groundwater by filtration. *Journal AWWA*, **59** (7), 878-896.

Gregory, R., Maloney, R.J. and Stockley, M. (1988) Water treatment using magnetite: A study of Sirofloc pilot plant. *Journal of IWEM*, **2**, October, 532-544.

Gu B., Schmitt, J., Chen, Z., Liang, L. and McCarthy J.F. (1994) Adsorption and desorption of natural organic matter on iron oxide: mechanisms and models. *Environmental Science and Technology*, **28** (1), 38-46.

Gu B., Schmitt, J. and McCarthy J.F. (1995) Adsorption and desorption of different organic matter fractions on iron oxide. *Geochimica et Cosmochimica Acta*, **59**, 219-225.

Hatva, T. (1988) Treatment of groundwater with slow sand filtration. *Water Science and Technology*, **20** (3), 141-147.

Hatva, T. (1989) *Iron and manganese in groundwater in Finland: Occurrence in glacifluvial aquifers and removal by biofiltration.* National Board of Waters and Environment, Helsinki, Finland.

Hauer, G.E. (1950) Iron and carbon dioxide removal. *Journal AWWA*, **42**, 555-561.

Hem, J.D. and Cropper, W.H. (1962) *Chemistry of Iron in Natural Waters.* US Geological Survey Water Supply Paper 1459, Washington DC.

Hem, J.D. (1989) *Study and interpretation of the chemical characteristics of natural water.* Third Edition, United States Geological Survey Water-Supply Paper 2254.

Home, G.P., Stockley, M.and Shaw, G. (1992) The Sirofloc process at Redmires water-treatment works. *Journal of IWEM*, **6**, February, 10-19.

Hughes, M.N. and Poole, R.K. (1989) *Metals and Microorganisms.* Chapman and Hall, London, UK.

Hult, A. (1973) Filtration of iron during and after oxidation. *Effluent Water Treatment Journal*, April, 209-215.

Ibrahim, A.K.M. (1997) *The development of iron oxide coating on filter media and its role in iron removal.* M.Sc. Thesis SEE 019, IHE, Delft, The Netherlands.

JICA (1990) *Groundwater Management Project in Kathmandu Valley: Final Report.* Japan International Cooperation Agency.

J.M.Montgomery Consulting Engineers (1985) *Water Treatment, Principles and Design.* John Wiley and Sons Inc., USA.

Jobin, R. and Ghosh, M.M. (1972) Effect of buffer intensity and organic matter on the oxygenation of ferrous iron. *Journal AWWA*, **64**, 590-595.

Joshi, A. and Chaudhuri, M. (1996) Removal of arsenic from groundwater by iron oxide coated sand. *Journal of Environmental Engineering - ASCE,* **122** (8), 769-771.

Katko, T.S.(1997) *Water! Evolution of Water and Sanitation in Finland from the mid-1800s to 2000.* Finnish Water and Waste Water Association; Hameen Kirjapaino Ltd., Tampere.

Knocke, W.R., van Benschoten, J.E., Kearney, M.J., Soborski, A.W. and Reckhow, D.A. (1991) Kinetics of manganese and iron oxidation by potassium permanganate and chlorine dioxide. *Journal AWWA*, **83** (6), 80-87.

Knocke, W.R., Conley, L. and van Benschoten, J.E. (1992) Impact of dissolved organic carbon on the removal of iron during water treatment. *Water Research*, **26** (11), 1515-1522.

Knocke, W.R., Shorney, H.L. and Bellamy, J.D. (1994) Examining the reaction between soluble iron, DOC and alternative oxidants during conventional treatment. *Journal AWWA*, **86** (1), 117-127.

Kothari, N. (1988) Groundwater, iron and manganese: An unwelcome trio. *Water/Engineering and Management*, February, 25-26.

Kruithof, J.C. and Koppers, H.M.M.(1989) Experiences with groundwater treatment and disposal of the eliminated substances in The Netherlands. *Journal of Water SRT- Aqua*, **38** (3), 207-216.

Lai, C.H., Lo, S.H. and Lin, C.F. (1994) Evaluating an iron oxide coated sand for removal of copper from water. *Water Science and Technology*, **30** (9), 175-182.

Lerk, C.F. 1965 *Enkele aspecten van de ontijzering van grondwater* (in Dutch). Ph.D. Dissertation, Technical University Delft, The Netherlands.

Lorenz, W., Seifert, K. and Kasch, O.K. (1988) Iron and manganese removal from a groundwater supply. *Public Works*, November, 57-58.

MOPE (2000) *The State of Population Nepal, 2000.* Ministry of Population and Environment, Kathmandu, Nepal.

Mouchet, P. (1992) From conventional to biological removal of iron and manganese in France. *Journal AWWA*, **84** (4), 158-167.

Nash, H. and McGall, G.J.H. (edited) (1994) *Groundwater Quality - 17th Special Report.* Chapman & Hall.

Oldham, W.K. and Gloyna, E.F. (1969) Effect of coloured organics on iron removal. *Journal AWWA*, **61** (11), 610-614.

Olson, L.L. and Twardowski Jr., C.J. (1975) $FeCO_3$ vs $Fe(OH)_3$ precipitation in water treatment plants. *Journal AWWA*, **67** (3), 150-153.

O'Connor, J.T. and Benson, B.E. (1970) Iron removal using magnesium oxide. *Journal of Sanitary Engineering Division- ASCE*, **96** (SA6), 1335-1348.

O'Connor, J.T. (1971) Iron and Manganese. In: *Water Quality and Treatment - A Handbook of Public Water Supplies.* Chapter 11, p 378-396; McGraw Hill Book Company, New York.

Paillard, H., Legube, B., Bourbigot, M.M. and Lefebvre, E. (1991) Iron and manganese removal with ozonation in the presence of humic substances. *Ozone Science and Engineering*, **11** (1), 93-114.

Raghunath, H.M. (1987) *Ground Water.* Second edition. Wiley Eastern Limited, New Delhi, India.

Rail, C.D. (1989) *Groundwater contamination: Sources, control and preventive measures.* Technomic Publishing Company, Inc., Pennsylvania, USA.

Robinson, L.R. Jr. and Breland, E.D. (1968) Removal of iron and manganese from low alkalinity waters. *Public Works*, **59** (2), 72-76.

Robinson, L.R. Jr. (1975) How silica affects iron removal from groundwater. *Water and Sewage Works*, **122** (3), 75-77.

Robinson, R.B., Demirel, T. and Baumann, E.R. (1981) Identity and character of iron precipitates. *Journal of Environmental Engineering - ASCE*, **107** (EE6), 1211-1227.

Rott, U. (1973*) Untersuchungen zur Aufbereitung von Grondwasser mit hohem Gehalt and Eisen und Huminsa'uren - Autokatalytische Enteisenung in Ein-aund Mehrschichtfiltern* (In German). Ph.D. Dissertation, Technical University of Hanover, Germany.

Rott, U. (1985) Physical, chemical and biological aspects of the removal of iron and manganese underground. *Water Supply*, **3** (2), 143-150.

Salvato, J.A. (1992) *Environmental Engineering and Sanitation.* Fourth Edition, John Wiley and Sons, Inc.

Sampat, P. (2000) Groundwater shock. *World Watch,* January/February, 10-22.
Sarikaya, H.Z. (1980) Interaction between ferrous iron oxidation and phosphate. In: *Treatment and Disposal of Liquid on Solid Wastes.* Pergamon Press, pp 143-160.

Schenk, J.E. and Weber Jr., W.J. (1968) Chemical interaction of dissolved silica with iron(II) and iron(III). *Journal AWWA*, **60** (2), 199-212.

Sharma, S.K. (1997) *Iron removal in water treatment by the sorption of iron(II) onto filter media.* M.Sc. Thesis SEE 018, IHE, Delft, The Netherlands.

Silver J.(1993) *Chemistry of Iron.* First Edition, Blackie Academie and Professional.

Singley, J.E. and Black, A.P. (1967) Hydrolysis products of iron(III). *Journal AWWA*, **59**, 1549-1564.

Singley, J.E. and Sullivan, J.H. (1969) Reactions of metal ions in dilute solution: Recalculation of hydrolysis of iron(II) data. *Journal AWWA*, **61,** 190-192.

Søgaard, E. G., Medenwaldt, R. and Abraham-Peskir, J.V. (2000) Conditions and rates of biotic and abiotic iron precipitation in selected Danish freshwater plants and microscopic analysis of precipitate morphology. *Water Research*, **34** (10), 2675-2682.

Sommerfeld, E.O. (1999) *Iron and Manganese Removal Handbook.* American Water Works Association, Denver, USA.

Stenkamp, V.S. and Benjamin, M.M. (1994) Effect of iron oxide coatings on sand filtration. *Journal AWWA*, **86** (8), 37-50.

Stevenson, D.G. (1997) *Water Treatment Unit Processes.* World Scientific Publishing Co. Pvt. Ltd. p 261-266, 275-293.

Stumm, W. and Morgan, J.J. (1996) *Aquatic Chemistry.* Third Edition, John Wiley & Sons Inc., USA.

Stumm, W. and Lee, G.F. (1961) Oxygenation of ferrous iron. *Industrial Engineering and Chemistry*, **53** (2), 143-146.

Sung, W. and Morgan, J.J. (1980) Kinetics and products of ferrous iron oxygenation in aquatic systems. *Environmental Science and Technology*, **14** (5), 561-568.

Tamura, H., Goto, K. and Nagayama, M. (1976) The effect of ferric hydroxide on the oxygenation of ferrous ions in neutral solutions. *Corrosion Science*, **16**, 197-207.

Theis, T.L., and Singer, P.C. (1973) The stabilization of ferrous iron in natural waters. In: *Trace Metals and Metal Organic Interaction in Natural Waters.* Edited by P.C. Singer, Ann Arbor Science Publishers, Michigan, USA.

Theis, T.L. and Singer , P.C. (1974) Complexation of iron(II) by organic matter and its effect on iron(II) oxygenation. *Environmental Science and Technology*, **8** (6), 569-573.

Tipping, E. (1981) The adsorption of aquatic humic substances by iron oxide. *Geochimica et Cosmochimica Acta*, **45**, 191-199.

Tufekci, N. and Sarikaya, H.Z. (1996) Catalytic effect of high Fe(III) concentration on Fe(II) oxidation. *Water Science and Technology*, **34** (7-8), 389-396.

Twort, A.C., Ratnayaka, D.D. and Brandt, M. J. (2000) *Water Supply.* Fifth edition, Arnold Publishing - Hodder Headline Group, London, UK.

UNICEF (1986) *Iron Removal Plant.* Water and Sanitation Section, UNICEF - Kathmandu Nepal.

van Beek, C.G.E.M. (1983) *In Situ Iron Removal -An Evaluation of Investigations*. Report 78 (In Dutch), Nieuwegein, KIWA N.V., The Netherlands.

van der Kooij, D., van Lieverloo, J.H.M., Schellert, J.A. and Hiemstra, P. (1999) Distributing drinking water without disinfectant: highest achievement or height of folly? *Journal of Water SRT-Aqua*, **48** (1), 31-37.

van Wijk, H.R., van Dijk, J.C. and de Moel, P.J. (1987) Filtratie van ijzerhoudend grondwater in Nederland (In Dutch). *H_2O*, **20** (6), 126-128.

VEWIN Recommendations 1993. In Dutch : VEWIN Aanbevelingen 1993 (a. de hoedanigheid van het water = waterkwalitiet), VEWIN, The Netherlands.

Vigneswaran, S. and Visvanathan, C. (1995) *Water Treatment Processes - Simple Options*. CRC Press, p 156-157.

WHO (1996) *Guidelines for drinking water quality: 2nd edition, Vol 2 - Health Criteria and other supporting information*. WHO, Geneva.

Willey, B.F. and Jennings, H. (1963) Iron and manganese removal with potassium permanganate. *Journal AWWA*, **55** (6), 729-734.

Wong, J.M. (1984) Chlorination-filtration for iron and manganese removal. *Journal AWWA*, **76** (1), 76-79.

Chapter 2

ADSORPTION OF IRON(II) ONTO FILTER MEDIA AND IRON HYDROXIDES

Part of this Chapter was published by S.K. Sharma, M.R. Greetham and J.C. Schippers (1999) *Journal of Water Supply: Research and Technology - Aqua,* **48** (3), 84-91.

ABSTRACT

Quantification of iron(II) adsorption capacities of filter media and iron hydroxides is essential for a better understanding and optimisation of the process of adsorptive iron removal in filters. A laboratory-scale batch experimental method was developed to measure the capacity of filter media and iron hydroxides to adsorb iron(II) ions. In addition to new (virgin) sand, iron oxide coated sand sampled from two groundwater treatment plants and other filter materials, namely anthracite, basalt, pumice, limestone, magnetite, and olivine were tested. All the adsorption experimental data fitted the Freundlich isotherm model. Among the virgin filter materials tested, basalt had the highest capacity for iron(II) adsorption followed by anthracite, olivine, magnetite, sand, pumice, and limestone. Compared to new filter sand, coated sand from the two groundwater treatment plants had much higher iron(II) adsorption capacity. At a pH of 6.5, their adsorption capacities were about 20 to 25-fold that of new sand. Iron(II) adsorption capacity of both new and iron oxide coated sand increased with the increase of pH. The iron(II) adsorption capacities of coated sand measured in model groundwater and real groundwater were different. Among the four common iron hydroxides tested, lepidocrocite had the highest iron(II) adsorption capacity, followed by amorphous iron hydroxide, ferrihydrite, and goethite. Iron(II) adsorption capacity of lepidocrocite decreased with ageing, probably due to its transformation to other mineralogical forms of iron oxides. Based on the experimental results, it was estimated that the contribution of iron(II) adsorption onto iron hydroxide flocs in adsorptive iron removal as well as floc filtration iron removal is likely very limited in iron removal plants.

2.1 INTRODUCTION

The adsorption-oxidation mechanism of iron removal relies on the iron(II) adsorption capacity of the filter media and its subsequent regeneration. The iron(II) adsorption capacity may depend on characteristics of the filter media, water quality and process conditions applied. Furthermore, iron(II) adsorption capacity of the new media increases with the development of the iron oxide coating. In the presence of oxygen in the raw water, some iron hydroxide flocs are likely to be formed. The characteristics of these iron hydroxides also depend on water quality and process conditions applied (Misawa *et al.* 1974; Robinson *et al.* 1981; Carlson and Schwertmann 1987; Cornell and Schwertmann 1996). Iron hydroxides can adsorb iron(II) and catalyse the iron oxidation (O'Connor 1971; Tamura *et al.* 1976), thereby enhancing the iron removal process. The adsorption capacity of iron hydroxides may also change with their ageing. In order to understand and optimise the process of adsorptive iron removal in filters, an accurate quantification of the iron(II) adsorption capacities of different filter media and iron hydroxides is required. Hence, the objectives of this study are:

i) to determine the iron(II) adsorption capacities of different new and iron oxide coated filter media and to analyse the effect of water quality, specifically pH, on iron(II) adsorption capacity and,

ii) to determine the iron(II) adsorption capacities of different iron hydroxides and to assess the effect of pH and ageing on iron(II) adsorption.

2.2 THEORETICAL BACKGROUND

2.2.1 Adsorption of iron(II) onto filter media

The adsorption of metal cations onto iron oxide has been well studied (Davis and Leckie 1978; Benjamin and Leckie 1981; Dzombak and Morel 1990). Davis and Leckie (1978) proposed a mechanism involving surface ionization and complexation of trace metal ions with hydrous iron oxide surfaces. In aqueous systems the surfaces of oxides are covered with surface hydroxyl groups. The acid-base equilibrium of a hydroxylated oxide surface is commonly represented as:

$$\equiv S\text{-}OH_2^+ \quad \equiv \quad \equiv S\text{-}OH + H^+ \qquad K_{a1} \tag{2.1a}$$

$$\equiv S\text{-}OH \quad \equiv \quad \equiv S\text{-}O^- + H^+ \qquad K_{a2} \tag{2.1b}$$

where $\equiv S\text{-}OH_2^+$, $\equiv S\text{-}OH$ and $\equiv S\text{-}O^-$ represent positively charged, neutral and negatively charged surface hydroxyl groups respectively and K_{a1} and K_{a2} are the acidity constants (Dzombak and Morel 1990).

$$K_{a1} = \frac{[\equiv S-OH][H^+]}{[\equiv S-OH_2^+]} \tag{2.2a}$$

$$K_{a2} = \frac{[\equiv S-O^-][H^+]}{[\equiv S-OH]} \tag{2.2b}$$

The adsorption of a metal ion on an oxide surface involves the formation of bonds of the metal ion with the surface oxygen atoms and the release of protons from the surface.

$$\equiv S\text{-}OH + M^{2+} \quad \equiv \quad \equiv S\text{-}OM^+ + H^+ \qquad K_M \tag{2.3}$$

where M^{2+} represents a divalent cation and K_M is surface complexation constant.

$$K_M = \frac{[\equiv S-OM^+][H^+]}{[\equiv S-OH][M^{2+}]} \tag{2.4}$$

There is also the possibility of forming the bidentate complexes.

$$\begin{matrix} \equiv S-OH \\ \equiv S-OH \end{matrix} + M^{2+} \rightarrow \begin{matrix} \equiv S-O \\ \equiv S-O \end{matrix}\rangle M + 2H^+ \qquad K_{M2} \tag{2.5}$$

$$K_{M2} = \frac{[(\equiv S-O)_2 M][H^+]^2}{[\equiv S-OH]^2[M^{2+}]} \tag{2.6}$$

Consistent with equation (2.3) or (2.5), the adsorption of cations onto the hydrous oxide surface increases with the increase in pH. However it has been found that for the given adsorbate/ adsorbent ratio, there is a narrow pH range, usually one or two units, over which the adsorption of cations on hydrous oxides increases from 0 to 100%, giving typical pH versus percentage adsorption curves known as adsorption "pH edges"(Davis and Leckie 1978; Benjamin and Leckie 1981; Dzombak and Morel 1990). As the adsorbate/adsorbent ratio is increased, the fractional adsorption at a given pH is reduced and consequently the cation "pH edges" shift to the right.

Adsorption is an important purifying action in deep bed filters. Adsorption can reduce soluble metal concentrations to well below the values achievable by precipitation. Adsorption can also remove inorganically and organically complexed metals (Cu, Cd, Pb, Zn) as well as oxyanions metals (SeO_3^{2-}, CrO_4^{2-}, AsO_4^{3-}) under conditions where conventional precipitation would be ineffective (Edwards and Benjamin 1989; Benjamin *et al.* 1996). Different factors may influence the adsorption of iron(II) onto filter media. The physicochemical nature of the filter media, especially the specific surface area and the surface chemistry (nature of the surface, zeta potential and charge characteristics), have profound effects on both rate and capacity of adsorption. The adsorption capacity of a solid adsorbent is generally proportional to the specific surface area (Faust and Aly 1983). Therefore, adsorption capacities of different filter media may vary considerably. Additionally, the surface characteristics of the media including surface charge and specific surface area change with the development of iron oxide coating on its surface. Hence a considerable difference in the iron(II) adsorption capacity of new and iron oxide coated sand can be expected. Furthermore, the presence of other ions and compounds in water may also influence the adsorption of the given adsorbate and their adsorption will likely alter the characteristics of adsorbent surface. Table 2.1 lists the various factors that influence the adsorption of an adsorbate onto the filter media.

Table 2.1 Factors affecting adsorption onto filter media

Characteristics of filter media	Nature of adsorbate	Water Quality	Others
1. Type of media	1. Solubility	1. pH	1. Filtration rate
2. Surface area	2. Surface charge	2. Other adsorbates	2. Rate of regeneration of adsorption sites
- Particle size	3. Molecular weight	3. Ionic concentration	
- Porosity	4. Size of the adsorbate molecule	4. Alkalinity	
- Density		5. Organic matter	
3. Chemistry of surface		6. Temperature	
- Surface charge			
- Surface site density			
4. Nature of the coating			
- Mineral form			
- Crystal structure			
- Time in use			

(Compiled from Weber 1972; Faust and Aly 1983; Dzombak and Morel 1990; Stumm and Morgan 1996)

2.2.2 Iron hydroxides and iron(II) adsorption

Iron (hydr)oxides exist in different mineralogical and morphological forms. There are sixteen different iron oxides, oxyhydroxides and hydroxides known to date (Cornell and Schwertmann 1996). Among them four iron hydroxides commonly found in the iron removal process are lepidocrocite (γ-FeOOH), amorphous iron(III) hydroxide ($Fe_2O_3.H_2O$), ferrihydrite ($Fe_5HO_8.4H_2O$), and goethite (α-FeOOH) (Tamura *et al.* 1976; Sung and Morgan 1980; Robinson *et al.* 1981; Carlson and Schwertmann 1987). The characteristics of the iron hydroxides formed depend on water quality and process conditions. Secondly, these iron oxides age over time to form stable end products. The occurrence, formation, characteristics and transformation pathways of different iron (hydr)oxides are well documented (Cornell *et al.* 1989; Schwertmann and Cornell 1991; Cornell and Schwertmann 1996).

Iron oxides are known to adsorb iron(II), however their iron(II) adsorption capacity may differ widely. Sung and Morgan (1980) observed that initial products of iron(II) oxidation are γ-FeOOH and amorphous FeOOH, and these forms are unstable and change over time to form α-FeOOH. Robinson *et al.* (1981) reported that γ-FeOOH was the first precipitate at pH <9 when oxygen was the oxidising agent and silica concentration was <7.1 mg/l as SiO_2 in the presence of 10 mg/l of iron. Mouchet (1992) mentioned that the precipitates formed during iron removal are slightly hydrated iron oxides, especially γ-FeOOH and sometimes α-FeOOH. Lepidocrocite is quickly replaced by ferrihydrite under increasing O_2 content, pH and Eh (Mouchet 1992).

It has also been reported that the presence of iron hydroxide accelerates the rate of iron(II) oxidation (Tamura *et al.* 1976; Tufekci and Sarikaya 1996, 1998). Takai (1973) studied the catalytic oxidation of iron(II) by α-FeOOH and γ-FeOOH in aqueous solution and found that γ-FeOOH was an effective catalyst for the oxidation of iron but that α-FeOOH had no effect. In contrast, Tufekci and Sarikaya (1998) reported that the catalytic effect of iron hydroxide on iron(II) oxidation increases with the ageing of the iron hydroxides. Hence, the iron hydroxide flocs in the filter may influence the process of iron removal by adsorbing iron(II) as well as by catalysing iron(II) oxidation.

2.2.3 Adsorption isotherm

In adsorption modelling, the distribution of solute between the liquid and the solid phase is often described in terms of isotherms. The presentation of the amount of solute adsorbed per unit of adsorbent as a function of the equilibrium concentration of solute in bulk solution, at constant temperature, is called the adsorption isotherm. The Langmuir and Freundlich models are the two commonly used adsorption isotherms in aqueous system (Weber 1972; Stumm and Morgan 1996). The Langmuir model is valid for single layer adsorption. It is based on the assumption that maximum adsorption corresponds to a saturated monolayer of solute on the adsorbent surface, the adsorption energy is the same at all sites and there is no transmigration of adsorbate in the plane of the surface.

The Freundlich isotherm is the most widely used mathematical description of adsorption in aqueous systems. It is based on the assumption of heterogeneous adsorption surface comprising of different classes of adsorption sites and energies. The Freundlich equation is an empirical expression but is often useful as a means of data description. Dzombak and Morel (1990) reported that cation sorption onto hydrous oxides can usually be approximated by the Freundlich isotherm. The Freundlich isotherm is mathematically represented as:

$$q = KC_e^n \tag{2.7}$$

where q = the amount of solute adsorbed per unit weight or surface area of the adsorbent (g/g or g/m^2), C_e = equilibrium concentration of the solute (g/m^3), K, n = isotherm constants. On linearisation, the equation takes the form

$$\log q = \log K + n \log C_e \tag{2.8}$$

Isotherm constant K is the measure of adsorption capacity and constant n is the measure of adsorption intensity. For fixed values of C_e and n, the higher the value of K, the higher is the adsorption capacity. For fixed values of K and C_e, the lower the value of n, the stronger is the adsorption bond. As n becomes very small, the capacity tends to be independent of C_e, and the isotherm plot approaches the horizontal. If the value of n is large, the adsorption bond is weak, and the value of q changes markedly with small changes in C_e (Snoeyink 1990).

2.3 MATERIALS AND METHODS

2.3.1 Filter materials

Iron(II) adsorption capacities of new (virgin) silica sand, anthracite, basalt, pumice, limestone, magnetite, olivine and naturally iron oxide coated sand from the filters of two groundwater treatment plants (Ridderkerk, WZHO and Gilze, WNWB) were determined. A sieve analysis of the media was conducted to determine the grain size distribution. The geometric surface area of the fractions retained on each sieve was calculated separately and the average geometric surface area per unit weight of the material was determined (Corstjens 1972; Huisman 1986). A sample calculation of the average geometric surface area of the filter media is presented in Appendix 2.1. Table 2.2 presents a size characterisation of the media tested.

The new sand was washed with demineralised water, soaked in concentrated hydrochloric acid (pH=1) for 24 hrs, again rinsed with demineralised water and dried at 40°C. The other virgin media were washed with demineralised water and dried at 40°C. The coated sands from the two water treatment plants were washed gently to remove any deposits and dried at room temperature.

Table 2.2 Size characterisation of the filter media analysed

Filter media	Grain size (mm)	Effective size d_{10} (mm)	Uniformity Coefficient d_{60}/d_{10}	Geometric surface area (m^2/g)
1. New sand	0.7 - 1.25	0.81	1.17	2.78×10^{-3}
2 Coated sand (Ridderkerk)	2.7 - 5.2	2.86	1.34	7.58×10^{-4}
3. Coated sand (Gilze)	0.7 - 1.25	0.77	1.23	2.82×10^{-3}
4. Anthracite	1.4 - 2.5	1.48	1.38	4.09×10^{-3}
5. Basalt	0.8 - 1.6	1.04	1.21	3.49×10^{-3}
6. Limestone	0.8 - 1.6	0.93	1.35	3.39×10^{-3}
7. Pumice	2.0 - 3.15	0.95	1.39	4.36×10^{-3}
8. Magnetite	0.5 - 0.8	0.53	1.25	3.17×10^{-3}
9. Olivine	0.6 - 1.6	0.65	1.52	2.57×10^{-3}

2.3.2 Iron hydroxides

Four different iron hydroxides namely lepidocrocite (γ-FeOOH), goethite (α-FeOOH), ferrihydrite ($Fe_5HO_8.4H_2O$) and amorphous iron(III) hydroxide ($Fe_2O_3.H_2O$) were tested for their iron(II) adsorption capacity. Amorphous iron hydroxide was prepared in solution by hydrolysing iron(III) nitrate ($Fe(NO_3)_3.9H_2O$) at pH 7.0 for 30 minutes (Tamura *et al.* 1976), its concentration being in the range 30- 60 mg iron(III)/l. Other iron hydroxides were prepared in the laboratory according to the method described by Schwertmann and Cornell (1991) and their mineralogy was confirmed by conducting powder X-ray diffraction of the samples using Siemens D500 Goniometer at the Laboratory of Material Science, Delft University of Technology.

2.3.3 Experimental set-up

A laboratory batch experimental set-up, as shown in Fig. 2.1, was developed to measure the adsorption of iron(II) onto filter media and iron hydroxides. The experimental apparatus consisted of a sealed 2-litre glass reactor. The vessel cover had ports to allow solution feeding and sampling, oxygen, temperature and pH measurement, gas supply, and mechanical stirring.

A model groundwater, prepared by mixing iron(II) sulphate ($FeSO_4 .7H_2O$) stock solution (400 mg/l, pH <2) with deoxygenated demineralised water, was used in the experiments. In adsorption experiments with filter media the pH was adjusted using the CO_2 /HCO_3^- buffer system, whereas in experiments with iron hydroxides the pH was adjusted using HCl and NaOH. 1.0 g/l of $NaHCO_3$ was dosed in the water for this purpose. All the chemicals used were of reagent grade. The experiments were conducted at room temperature (18-20°C).

Nitrogen gas was passed through the water to strip out the oxygen. The bubbling of nitrogen was continued throughout the experiment to control the oxygen concentration and pH. The pH and oxygen concentration were monitored continuously with a pH meter (WTW pH 323) and oximeter (WTW Oxi 340) to ensure constant pH and anoxic conditions. The gas provided the positive pressure against the inflow of air. By maintaining anoxic conditions, iron was kept in the iron(II) form. Consequently, the adsorbed iron(II) could not be oxidised and the adsorption sites could not be regenerated.

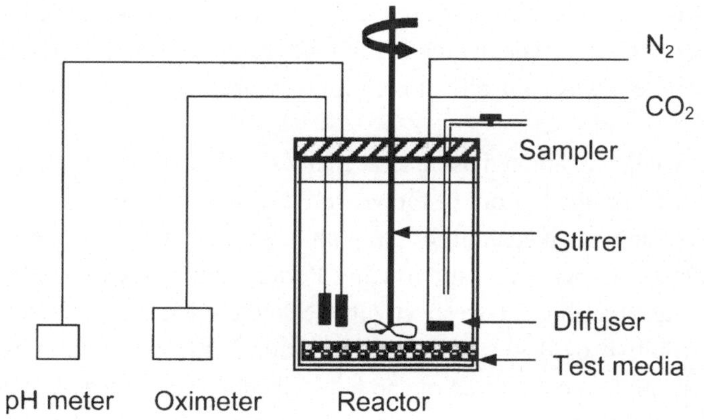

Fig. 2.1a Schematic of batch experimental set-up used for iron(II) adsorption measurement

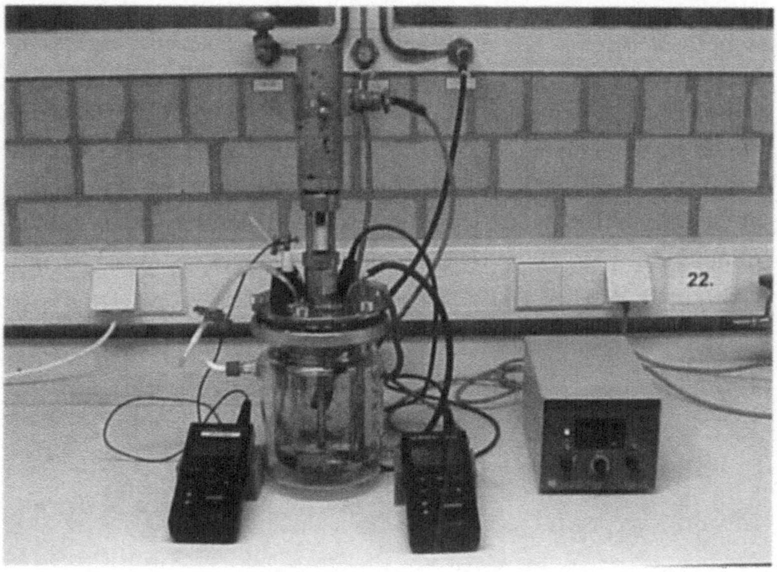

Fig. 2.1b Photo of the batch experimental set-up

2.3.4 Experimental procedure

Initially, 1.5 litres of demineralised water was deoxygenated by bubbling nitrogen gas through the solution and then the pH was adjusted to the desired level. Deoxygenated iron(II) stock solution (400 mg/l, pH <2) was then dosed into the reactor containing a known amount of filter material or iron hydroxides. An initial pH drop of up to 0.2 units was observed on addition of the acidified iron(II) stock solution. The pH returned to the required value within 15-25 minutes. The pH during the experiment was controlled to within 0.05 units.

The solution in the reactor was mixed using a mechanical stirrer. Throughout the experiment, the media remained settled on the bottom of the vessel, thus avoiding any scouring of the surface coating. When the system attained equilibrium with respect to iron(II) adsorption, the total iron and iron(II) concentrations were measured using Perkin Elmer AAS 3110 and Perkin Elmer UV/VIS Spectrometer Lambda 20 respectively (Standard Methods 1995). In the case of experiments with iron hydroxides, iron(II) concentration in the presence of excess of iron(III) was measured according to the modified method described by Tamura *et al.* (1974). The amount of iron(II) adsorbed onto the filter media or iron hydroxide was then calculated from an iron mass balance. Blank experiments without filter media or iron hydroxide revealed that the loss of iron(II) due to adsorption onto the reactor was <1% of the total iron.

The equilibration time for each filter media or iron hydroxide was determined separately by conducting trial experiments and measuring the hourly variation in iron(II) concentration. It was found that most of the adsorption (about 80%) on the filter media takes place within the first two hours. Equilibrium was considered as being reached when the difference in the iron(II) concentrations of the two subsequent hourly samples were <0.05 mg/l. The weight of the filter material used in the experiments was varied to keep the time needed to reach equilibrium within 4-8 hours. The weights of the media tested and corresponding equilibration times are presented in Table 2.3. In the case of iron hydroxides, the adsorption was very rapid and the equilibration time was 60-75 minutes.

Table 2.3 Equilibration time of the filter materials

Filter Media	Equilibration time (hour)	Weight of the media tested (g)
1. New sand	4	100
2. Coated sand (Ridderkek)	4	100
3. Coated sand (Gilze)	6	50
4. Anthracite	7	25
5. Basalt	6	75
6. Limestone	5	50
7. Pumice	5	25
8. Magnetite	5	40
9. Olivine	6	50

The amounts of iron(II) adsorbed per unit geometric surface area of filter media at different equilibrium concentrations were plotted on a log-log graph. The Freundlich isotherm constants K and n were calculated by linear regression. For iron hydroxides, the iron(II) adsorption results were expressed in terms of iron(II) adsorbed per unit weight of iron(III).

2.4 RESULTS

2.4.1 Adsorption isotherms of filter media

The adsorption capacities of the different filter media for iron(II) were measured by conducting the batch adsorption equilibrium experiments in duplicates. The experimental adsorption data for all filter media tested fitted the Freundlich isotherm model corresponding to straight lines with slopes less than 1.0 on log-log graphs. This is in agreement with the observations of other researchers studying the adsorption of cations onto hydrous oxide surfaces (Benjamin and Leckie 1981; Dzombak and Morel 1990). Table 2.4 presents the Freundlich isotherm constants derived for different media.

Table 2.4 Freundlich isotherm constants for iron(II) adsorption onto filter media

Filter Material	pH	K $[g/m^2/(g/m^3)^n]$	n	r^2
New sand	6.0	2.32×10^{-3}	0.54	0.95
	6.5	2.48×10^{-3}	0.73	0.98
	7.0	3.85×10^{-3}	0.70	0.98
	7.5	11.6×10^{-3}	0.45	0.94
Coated sand (Ridderkerk)	6.5	48.0×10^{-3}	0.60	0.99
Coated sand (Gilze)	6.0	42.7×10^{-3}	0.49	0.96
	6.5	61.1×10^{-3}	0.54	0.99
	7.0	117.0×10^{-3}	0.43	0.98
	7.5	168.0×10^{-3}	0.52	0.97
	6.5*	57.4×10^{-3}	0.48	0.98
Anthracite	6.5	7.90×10^{-3}	0.78	0.97
Basalt	6.5	18.1×10^{-3}	0.38	0.99
Limestone	6.5	0.54×10^{-3}	0.60	0.90
Pumice	6.5	1.54×10^{-3}	0.94	0.92
Magnetite	6.5	2.88×10^{-3}	0.63	0.92
Olivine	6.5	6.46×10^{-3}	0.52	0.91

* Experiment conducted in the treated water from Gilze groundwater treatment plant

Figure 2.2 compares the adsorption capacities for iron(II) of different filter materials tested at pH 6.5. It is clear from Fig. 2.2 that iron(II) adsorption capacities of iron oxide coated sand are much higher than that of new (virgin) filter materials. At a pH of 6.5, the isotherm constants K for coated sand from Gilze water treatment plant (WTP) and Ridderkerk WTP were respectively about 25 and 20-fold that of new sand. Among the media tested, the coated sand from the Gilze WTP had the highest and limestone the lowest capacity for iron(II) adsorption. The difference in adsorption capacity of various media are likely due to the difference in the surface charge, the density of adsorption sites, and the mineralogy (compositions and crystal structures) of the media surfaces (Weber 1972; Stumm and Morgan 1996). In general, a higher iron(II) adsorption capacity was observed in virgin media like basalt and olivine with iron-bearing minerals and in iron oxide coated media. Therefore, the increased iron(II) adsorption capacity is likely due to the presence of iron oxides on the surface.

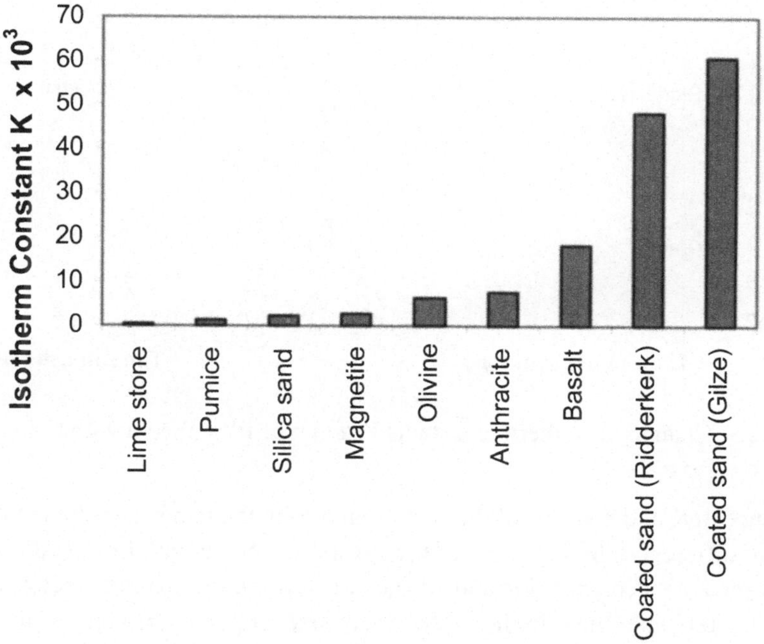

Fig. 2.2 Isotherm constant K [in $g/m^2/(g/m^3)^n$] for iron(II) adsorption onto different filter materials (pH = 6.5)

2.4.2 Iron(II) adsorption capacity of coated sand

The results of batch adsorption experiments clearly demonstrated the effectiveness of the iron oxide coating on adsorbing iron(II). The sand from the Gilze WTP was in use for three and a half years. The average influent iron concentration was 1.7 mg/l and the pH 7.6. The sand from the dry filter of Ridderkerk WTP was in use for two years. The average influent iron concentration was 7 mg/l and the pH 7.1. The sand from the Gilze WTP was fully coated with a dark brown iron oxide, while the sand from the Ridderkerk WTP was less coloured and only partially covered with iron oxide. Figure 2.3 shows the variation of the isotherm constants K and n of the sands from these two plants as well as of new sand at a pH of 6.5. The isotherm constant K for coated sand from Gilze WTP was about 25-fold that of new sand and 1.3-fold that of coated sand from Ridderkerk WTP. Figure 2.3 indicates that filter sand gets coated with iron oxide over time and hence the isotherm constant K increases. The isotherm constant n apparently decreased with time in use. In comparison to new sand, the adsorption capacity of iron oxide coated sand was less influenced by the iron(II) concentration as n is smaller. This indicates that iron(II) adsorption sites on coated sand are relatively less concentration-dependent than the adsorption sites on new sand.

Fig. 2.3 Change in isotherm constants K and n of filter media with time in use

The difference in K and n values of the coated sand from the two water treatment plants is very likely due to differences in the extent of the surface coverage with iron oxide coating and the mineralogy (chemical composition and crystal structure) of the coating. Factors affecting these properties include the time in use, the grain size and the influent water quality (iron concentration, pH, alkalinity, organic matter content, etc.). Additionally, the operational modes of the filter (dry or wet/normal), filtration rate and backwashing conditions may also influence the characteristics of the naturally coated sand. Benjamin *et al.* (1996) reported that the method of coating the filter media influences the surface characteristics of the coating (specific surface area, point of zero charge and zeta potential) and its adsorption capacity. In addition, Stenkamp and Benjamin (1994) concluded that the nature of the underlying media could influence the adsorption characteristics of the media coating, particularly when the coating is thin. This implies that iron(II) adsorption capacity of iron oxide coated sand developed under different water quality and process conditions applied at iron removal plants may vary considerably. Characteristics of iron oxide coated sand from different iron removal plants and their iron(II) adsorption capacities are further discussed in Chapter 4.

2.4.3 Effect of pH on iron(II) adsorption onto filter media

Figure 2.4 presents the adsorption isotherms for new and coated sand from the Gilze WTP at different pH. For both media, adsorption capacity increased with an increase of pH. The point of zero charge (pzc) for silica is approximately 2 (Stumm and Morgan 1996). In the pH range investigated (6.0 to 7.5), the silica surface is negatively charged, consequently having a high affinity for positively charged ions. As the pH increases the surface charge becomes more negative, therefore the adsorption capacity increases. For both media, the slopes of the isotherms at different pH are approximately identical. This indicates that the increase in the adsorption capacity with the increase of equilibrium concentration is nearly the same at the pH range investigated.

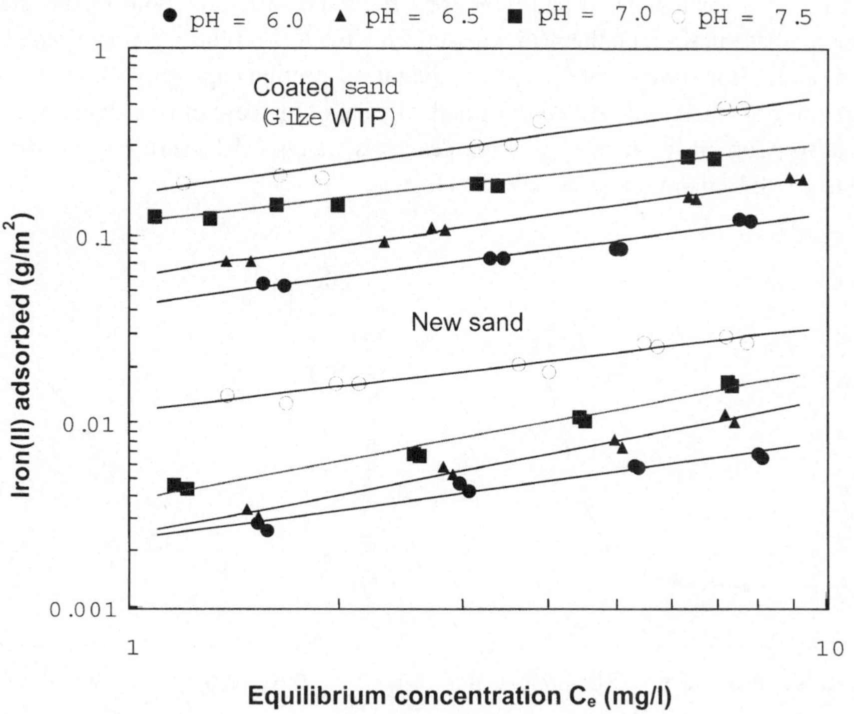

Fig. 2.4 Adsorption isotherms for iron(II) adsorption onto new sand and coated sand in demineralised water

The sand from the Gilze WTP had developed iron oxide coating on its surface during the normal operation of the iron removal process. The point of zero charge for hydrous ferric oxides (HFO) is between 7.8 - 8.2 depending on the ionic concentration (Dzombak and Morel 1990; Bailey *et al.* 1992). In the pH range 6.0 - 7.5, the iron oxide coating on the sand is positively charged, hence the adsorption of iron(II) takes place against the electrostatic repulsion. It is to be noted that adsorption of cations onto surfaces is a result of two major forces; the chemical forces or the short range forces, and the coulombic forces otherwise known as long range forces (Dzombak and Morel 1990; Stenkamp and Benjamin 1994; Stumm and Morgan 1996). Therefore, it is likely that chemical forces are more dominant in cases of iron(II) adsorption onto iron oxide coated sand. As the pH increases the hydrated oxide surface becomes less positively charged. There are more sites available for iron(II) adsorption, hence the adsorption capacity increases (Dzombak and Morel 1990; Stumm and Morgan 1996). The increase of adsorption with pH is consistent with acid-base equilibria of hydrated oxide surfaces as represented by equations (2.1) to (2.6).

The increase of the isotherm constant K with pH for new and coated sand is illustrated in Fig. 2.5. The constant K for new sand increased only by 7% when pH increased from 6.0 to 6.5, whereas for the coated sand the increase was 43%. With an increase in pH from 6.5 to 7.0, K increased by 55% for new sand and 91% for the coated sand. For both media there was a significant increase in K when the pH increased from 6.0 to 7.5 (400% for new sand and 293% for coated sand). This shows that for both media the adsorption capacity (isotherm constant K) is very sensitive to changes in pH. As expected, the rate of increase of K with pH was not linear since the adsorption of cation on hydrous oxides gives the typical "adsorption pH edges" (Davis and Leckie 1978; Benjamin and Leckie 1981).

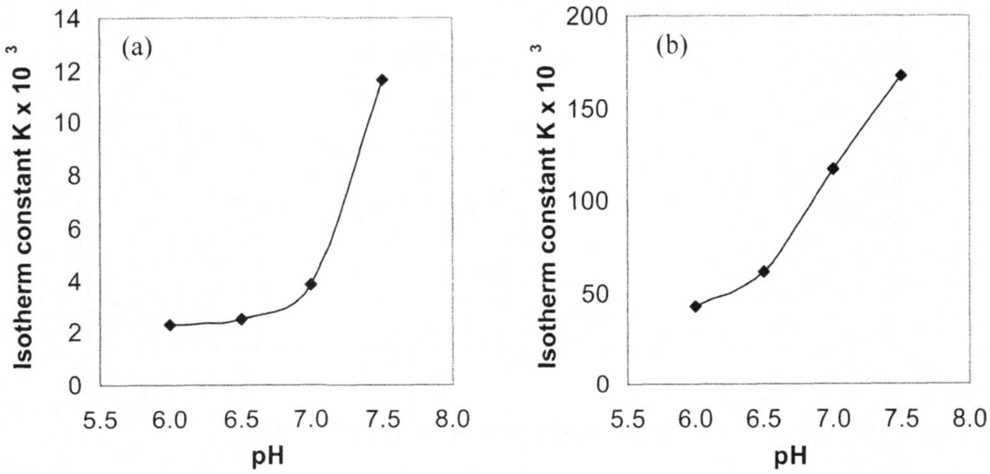

Fig. 2.5 Increase of isotherm constant K with pH for (a) new sand and (b) coated sand from Gilze WTP.

2.4.4 Effect of water quality on adsorption capacity of the filter media

In order to make a preliminary assessment of the effect of water quality on iron(II) adsorption onto filter media, the iron(II) adsorption capacity of coated sand from Gilze WTP was also measured in the treated groundwater from Gilze WTP spiked with iron(II) sulphate to get the raw water composition. Experiments were conducted as before at pH 6.5 ± 0.05 but without adding $NaHCO_3$ in the water. In this case, the pH was controlled utilising the natural buffer capacity of the water. The results were then compared with the iron(II) adsorption capacity of coated sand from Gilze WTP measured earlier in the model groundwater.

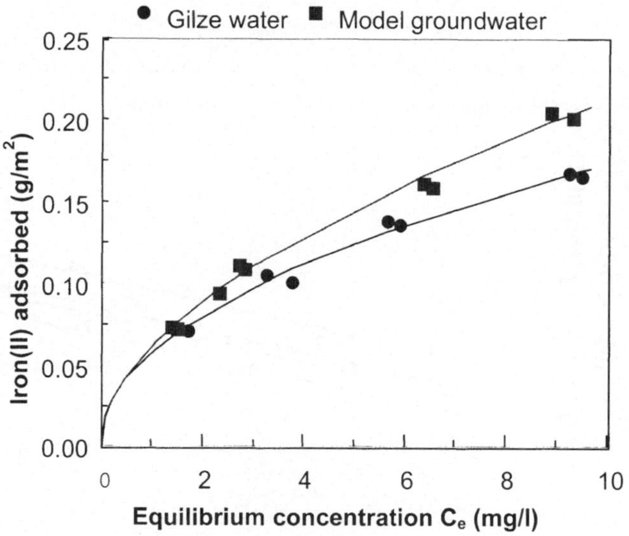

Fig. 2.6 Effect of water quality (type) on iron(II) adsorption onto coated sand from Gilze WTP

Figure 2.6 shows the iron(II) adsorption in the actual groundwater to be lower than that in the model groundwater. The isotherm constant K for iron(II) adsorption measured in treated water from Gilze WTP was about 6% lower than that in the model groundwater. This may be because of the presence of different ions and compounds in the actual groundwater. The TOC content of Gilze water (1.5 mg/l) is higher than that of the model groundwater (0.4 mg/l) and it is likely that the characteristics of the organic matter present are different. Organic matter could form complexes with iron(II) and iron(III) (Theis and Singer 1974) and could adsorb on the iron oxide coating, masking the properties of the underlying surface (Davis and Leckie 1978; Tipping 1981; Gu *et al.* 1994). The groundwater from Gilze WTP also contains different ions like calcium (52 mg/l), chloride (15 mg/l) and sulphate (13 mg/l) that may enhance, act relatively independently, or interfere with iron(II) adsorption (Benjamin 1983; Gu *et al.* 1994). The ions present in the water might alter the surface charge density and the thickness of the double layer. In this case the overall effect of these ions and compounds was to lower the iron(II) adsorption capacity of the media. Results obtained demonstrated the possible effect of water quality on iron(II) adsorption. The effect of water quality parameters on iron(II) adsorption onto new and iron oxide coated sand is further discussed in Chapter 3.

2.4.5 Adsorption of iron(II) onto different iron hydroxides

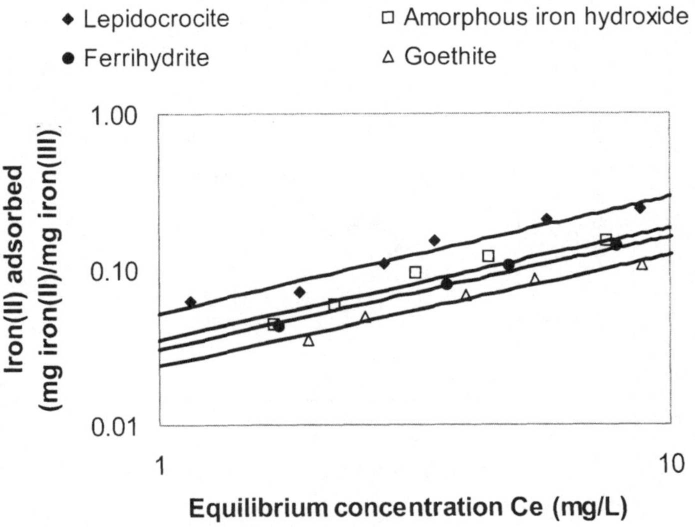

Fig. 2.7 Freundlich isotherms of iron(II) adsorption onto different iron hydroxides (iron(III) = 30 mg/l, pH = 7.00 ± 0.05, alkalinity = 0.01mol/l HCO_3^-).

Figure 2.7 presents the Freundlich adsorption isotherm for different iron hydroxides. The related isotherm constants are summarised in Table 2.5. Among the iron hydroxides tested, lepidocrocite had the highest iron(II) adsorption capacity, followed by amorphous iron hydroxide, ferrihydrite and goethite. According to Crosby *et al.* (1983), the surface area and the porosity of oxyhydroxides could contribute to their different adsorption capacities. Lepidocrocite has a surface area of 70-80 m^2/g as compared to goethite with approximately 20 m^2/g (Schwertmann and Cornell 1991). Therefore, lepidocrocite is expected to have a higher adsorption capacity than goethite. The difference in adsorption capacity of amorphous iron hydroxide and ferrihydrite is very small. This may be because both the oxides are amorphous and have almost similar structures.

Table 2.5 Freundlich isotherm constants for iron(II) adsorption onto iron hydroxides

Iron hydroxide	K (mg/mg/(mg/l)^n)	n	r^2
Lepidocrocite	5.22×10^{-2}	0.75	0.96
Amorphous	3.54×10^{-2}	0.71	0.93
Ferrihydrite	3.09×10^{-2}	0.71	0.98
Goethite	2.39×10^{-2}	0.71	0.98

2.4.6 Effect of pH on iron(II) adsorption onto iron hydroxide

Iron(II) adsorption onto amorphous iron hydroxide was measured by conducting batch experiments at initial iron(II) concentration of 6.0 mg/l and at different pH values, to assess the effect of pH on iron(II) adsorption onto iron hydroxides. Table 2.6 shows that iron(II) adsorption onto iron hydroxide also increased with the increase in pH.

Table 2.6 Increase in percentage iron(II) adsorption onto amorphous iron hydroxide with pH (iron(III) = 60 mg/l, initial iron(II) concentration = 6.0 mg/l, alkalinity = 0.01mol/l HCO_3^-).

pH	5.5	6.0	6.5	7.0	7.5
Iron(II) adsorbed (%)	12	23	43	71	99

As discussed in the case of filter media (section 2.4.3), the increased adsorption at higher pH is due to the pH and the surface charge effect. The higher the pH, the more easily the protons of the ≡Fe-OH group can be released, creating a negatively charged ≡Fe-O$^-$ group. This favours the adsorption of the positively charged Fe^{2+} ions. Secondly, as the pH increases to the point of zero charge of iron (hydr)oxides, the surface charge of the iron hydroxide gradually changes from more positive to less positive and finally to negative and this favours adsorption of iron(II).

2.4.7 Effect of lepidocrocite ageing on iron(II) adsorption

The decrease in the adsorption of iron(II) onto lepidocrocite with its ageing is shown in Fig. 2.8. Iron(II) adsorption after 10, 20 and 30 days was reduced to 89%, 84% and 80% of the adsorption on the first day respectively. The adsorption capacity of lepidocrocite on the thirtieth day was 0.08 mg iron(II)/mg iron(III) at the equilibrium iron(II) concentration of 3.5 mg/l. For freshly prepared lepidocrocite at the same equilibrium concentration, the iron(II) adsorption capacity was 0.13 mg iron(II)/mg iron(III). This shows that the adsorption capacity of lepidocrocite had reduced by about 40% in thirty days. It is expected that the decrease in iron(II) adsorption capacity would continue until the lepidocrocite is completely transformed to goethite, which has a capacity of about 0.06 mg iron(II)/mg iron(III). This clearly shows that ageing of iron hydroxide has a strong influence on its iron(II) adsorption capacity.

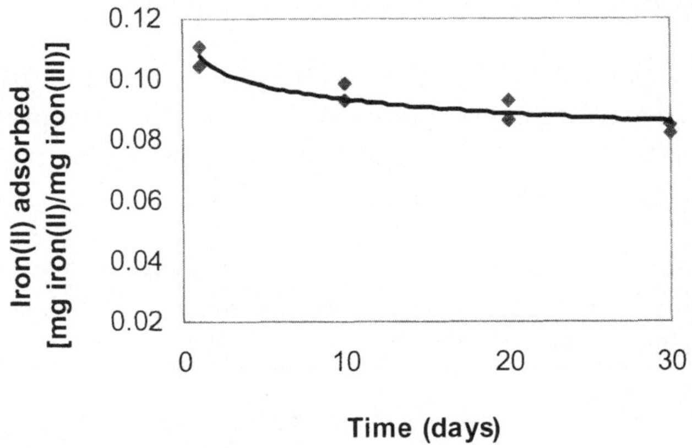

Fig. 2.8 Effect of lepidocrocite ageing on adsorption of iron(II) (iron(III) = 30 mg/l, initial iron(II) concentration = 6.0 mg/l, pH = 7.00 ± 0.05, alkalinity = 0.01mol/l HCO_3^-).

Crosby *et al.* (1983) reported that ageing affects both the surface area and the pore size of the iron hydroxides and this change is indicative of the changes in the structure. Lepidocrocite undergoes continuous transformation to goethite, which is more stable (Sung and Morgan 1980; Robinson *et al.* 1981; Schwertmann and Cornell 1991). The transformation of lepidocrocite (coupled with the reduction in surface area per unit weight) to other forms like goethite or hematite with time may be responsible for the decrease in iron(II) adsorption capacity.

2.5 DISCUSSION

Experimental results demonstrated that the different filter media tested can adsorb iron(II) on its surface and the adsorption capacity increases with pH and varies between media. The iron(II) adsorption capacity is greatly increased with the development of iron oxide coating on the media. This high iron(II) adsorption capacity of the coated media can be utilised to remove iron(II) from groundwater, provided that filters are operated predominantly in adsorptive filtration mode. Therefore, maintenance of proper process conditions is necessary to exploit the iron(II) adsorption capacities of filter media for iron removal. As discussed in Chapter 1, the predominantly adsorptive iron removal in filters can be achieved in two operational modes: i) in intermittent regeneration mode and ii) in continuous regeneration mode.

It should be noted that filter media may have a high iron(II) adsorption capacity, however the adsorbed iron(II) needs to be oxidised quickly to regain the "catalyst" or the adsorption capacity. The rate of oxidation of adsorbed iron(II) is likely to be much higher than that of iron(II) in water. This may also be influenced by water quality and process conditions. Hence, understanding of the rate and factors affecting the regeneration of adsorption sites is equally important, which needs a detailed study.

Contribution of iron(II) adsorption onto iron hydroxide flocs in iron removal

In the continuous regeneration mode of adsorptive iron removal or in the presence of traces of oxygen, some iron hydroxide flocs will very likely be formed. Consequently, some iron(II) could also be adsorbed onto these iron hydroxide flocs. An estimation of the iron(II) adsorption onto iron hydroxide flocs in an iron removal filter was made to assess the contribution and/or importance of the iron hydroxide flocs in adsorptive iron removal. For influent iron concentration of 5 mg/l, filtration rate of 8 m/h and run time of 48 hours, calculations (see Appendix 2.2) show that when 10% of iron(II) is oxidised to form flocs, the contribution of the flocs is less than 1%. Even when 50% of iron(II) is oxidised, due to high amounts of oxygen present or longer detention time, the average concentration of the flocs will be 2.5 mg iron(III)/l and their contribution to overall iron(II) adsorption will be about 2.5%. These calculations indicate that the contribution of the iron(II) adsorption onto iron hydroxide flocs in adsorptive iron removal as well as floc filtration iron removal is very limited in iron removal plants.

2.6 CONCLUSIONS

- Adsorption of iron(II) onto filter media and iron hydroxides fitted the Freundlich isotherm model with $n < 1.0$ and the isotherm constant K was used as a measure of the iron(II) adsorption capacity.
- Of the virgin materials tested, basalt showed the highest iron(II) adsorption capacity followed by anthracite, olivine, magnetite, sand, pumice and limestone.
- Iron oxide coated sands from full-scale iron removal plants demonstrated a much higher capacity for iron(II) adsorption compared to new filter media. Adsorption capacities of coated media from two water treatment plants were different, likely due to the difference in the extent of iron oxide coating, the time in use, water quality and the treatment processes employed.
- The iron(II) adsorption capacity of both new (virgin) and iron oxide coated sand increased with the increase of pH.
- There was some difference in the iron(II) adsorption capacity of coated sand measured in the model groundwater and the real groundwater. This indicates that different ions and compounds present in water could influence iron(II) adsorption, which needs further study.
- Among the iron hydroxides tested, lepidocrocite had the highest iron(II) adsorption capacity, followed by amorphous iron hydroxide, ferrihydrite and goethite. The iron(II) adsorption capacity of iron hydroxide also increased with pH. The iron(II) adsorption capacity of lepidocrocite decreased with ageing, probably due to its transformation to other mineralogical forms of iron oxides.
- An estimation based on the experimental results indicated that the contribution of iron(II) adsorption onto iron hydroxide flocs in the overall process of adsorptive iron removal as well as floc filtration iron removal is likely negligible in iron removal plants.

REFERENCES

Bailey, R.P., Bennett, T. and Benjamin, M.M. (1992) Sorption onto and recovery of Cr(VI) using iron oxide coated sand. *Water Science and Technology*, **26** (5-6), 1239-1244.

Benjamin, M.M. (1983) Adsorption and surface precipitation of metals on amorphous iron oxyhydroxides. *Environmental Science and Technology*, **17** (11), 686-691.

Benjamin, M.M. and Leckie, J.O. (1981) Multiple site adsorption of Cd, Cu, Zn and Pb on iron oxyhydroxide. *Journal of Colloid and Interface Science*, **79** (1), 209-221.

Benjamin, M.M., Sletten, R.S., Bailey, R.P. and Bennett, T. (1996) Sorption and filtration of metals using iron oxide coated sand. *Water Research*, **30** (11), 2609-2620.

Carlson, L. and Schwertmann, U. (1987) Iron and manganese oxides in Finnish groundwater treatment plants. *Water Research*, **21** (2), 165-170.

Cornell, R.M. and Giovanoli, R. and Schneider, W. (1989) Review of the hydrolysis of iron(III) and the crystallization of amorphous iron(III) hydroxide hydrate. *Journal of Chemical Technology and Biotechnology*, **46**, 115-134.

Cornell, R.M. and Schwertmann, U. (1996) *The iron oxides- structure, properties, reactions, occurrence and uses.* Published jointly by VCH- Germany and USA.

Corstjens, G.H. (1972) Berekening van opwaartse doorstroming en het terugspoelen van filterbedden (in Dutch). *H_2O*, **5** (25) 599-606.

Crosby, S. A. *et al.* (1983) Surface areas and porosities of Fe(III)- and Fe(II)-derived oxyhydroxides. *Environmental Science and Technology*, **17** (12), 709-713.

Davis, J.A. and Leckie, J.O. (1978) Surface ionization and complexation at the oxide/water interface: Surface properties of amorphous iron oxyhydroxide and adsorption of metal ions. *Journal of Colloid and Interface Science*, **67** (1), 90-107.

Dzombak, D.A. and Morel, F.M.M. (1990) *Surface Complexation Modelling - Hydrous Ferric Oxides.* John Wiley & Sons Inc., USA.

Edwards, M. and Benjamin, M.M. (1989) Adsorptive filtration using coated sand a new approach for the treatment of metal bearing wastes. *Journal of WPCF*, **61** (9), 1523-1533.

Faust, S.D. and Aly, O.M. (1983) *Chemistry of Water Treatment.* Ann Arbor Science Book.

Gu, B., Schmitt, J., Chen, Z., Liang, L. and McCarthy J.F. (1994) Adsorption and desorption of natural organic matter on iron oxide: Mechanisms and models. *Environmental Science and Technology*, **28** (1), 38-46.

Huisman, I. (1986) *Rapid Filtration.* IHE Lecture notes EE178/86/1, IHE Delft, The Netherlands.

Misawa, T., Hashimoto, K. and Shimodaira, S. (1974) The mechanisms of formation of iron oxide and oxyhydroxides in aqueous solutions at room temperature. *Corrosion Science*, **14**, 131-149.

Mouchet, P. (1992) From conventional to biological removal of iron and manganese in France. *Journal AWWA*, **84** (4), 158-167.

O'Connor, J.T. (1971) Iron and Manganese. In: *Water Quality and Treatment - A Handbook of Public Water Supplies.* McGraw Hill Book Company, New York, Chapter 11, pp 378-396.

Robinson, R.B., Demirel, T. and Baumann, E.R. (1981) Identity and character of iron precipitates. *Journal of Environmental Engineering Division* - ASCE, **107** (EE6), December, 1211-1227.

Schwertmann, U. and Cornell, R.M. (1991) *Iron Oxides in the Laboratory: Preparation and Characterisation.* Published jointly by VCH- Germany and USA.

Snoeyink, V.L. (1990) Adsorption of Organic Compounds. In: *Water Quality and Treatment - A Handbook of Community Water Supplies.* McGraw Hill Inc., Chapter 13, pp 781-875.

Standard Methods for the Examination of Water and Wastewater (1995) 19th edition, American Public Health Association / American Water Works Association /Water Environment Federation, Washington, DC.

Stenkamp, V.S. and Benjamin, M.M. (1994) Effect of iron oxide coatings on sand filtration. *Journal AWWA*, **86** (8), 37-50.

Stumm, W. and Morgan, J.J. (1996) *Aquatic Chemistry.* Third Edition, John Wiley & Sons Inc., USA.

Sung, W. and Morgan, J.J. (1980) Kinetics and products of ferrous iron oxygenation in aquatic systems. *Environmental Science and Technology*, **14** (5), 561-568.

Takai, T. (1973) Studies on the mechanisms of catalytic deferrization (II) - Determination of chemical structure of the deferrizing catalyst. *Journal of Japan Water Works Association*, No 466, July, 22-33.

Tamura, H. Goto, K. Yotsuyanagi, T. and Nagayama, M. (1974) Spectrophotometric determination of iron(II) with 1,10-phenanthroline in the presence of large amounts of iron(III). Short Communications, *Talanta*, **21**, 314-318.

Tamura, H., Goto, K. and Nagayama, M. (1976) The effect of ferric hydroxide on the oxygenation of ferrous ions in neutral solutions. *Corrosion Science*, **16**, 197-207.

Theis, T.L. and Singer, P.C. (1974) Complexation of iron(II) by organic matter and its effect on iron(II) oxygenation. *Environmental Science and Technology*, **8** (6), 569-573.

Tipping, E. (1981) The adsorption of aquatic humic substances by iron oxide. *Geochim. Cosmochim. Acta*, **45**, 191-199.

Tufekci, N. and Sarikaya, H.Z. (1998) Influence of ageing on the catalytic activity of ferric sludge for oxidation of Fe(II). *Water Science and Technology*, **38** (6), 129-137.

Tufekci, N. and Sarikaya, H.Z. (1996) Catalytic effect of high Fe(III) concentration on Fe(II) oxidation. *Water Science and Technology*, **34** (7-8), 389-396.

Weber, W.J. Jr. (1972) *Physicochemical processes for water quality control.* Wiley Interscience Publication; John Wiley & Sons Inc., USA.

APPENDIX 2.1 Calculations of average geometric surface area of filter grains

For uniform spherical grains of diameter d the total surface area per unit weight of filter media is given by

$$S = \frac{6}{d\rho} \tag{2.A.1}$$

where S = total surface area per unit weight (g/m^2)
d = diameter of filter grain (mm)
Δ = density of filter media (Kg/m^3)

Square woven sieves pass spherical grains of diameter d = s, giving the surface area of a fraction between the consecutive sieve s_i to s_j by

$$S = \frac{6}{\sqrt{s_i s_j}\,\rho} \tag{2.A.2}$$

Non-spherical grains have on one hand a larger surface to weight ratio, while the square woven sieves may also pass elongated grains with a volume in excess of Bs^3/6. Both factors are taken into account simultaneously by a shape factor N, giving the surface area per unit weight of the fraction between the sieve sizes s_i to s_j by

$$S = \frac{6}{\phi\sqrt{s_i s_j}\,\rho} \tag{2.A.3}$$

When a sample of filter media of weight W contains n fraction of weights W_1 to W_n between the consecutive sieves of sizes s_1 to s_{n+1}, the average geometric surface is given by the weighted average of the surface areas of the fractions retained on each sieve (Corstjens 1972; Huisman 1986). Table 2.A1.1 summarises a sample calculation of average geometric surface area of the new silica sand.

Table 2.A.1 Calculation of the average geometric surface area of the new silica sand

Lower sieve s_i (mm)	1.25	1.12	1	0.9	0.8	0.63	Total
Upper sieve s_j (mm)	1.4	1.25	1.12	1	0.9	0.8	
Mean size (mm)	1.32	1.18	1.06	0.95	0.85	0.71	
Shape factor ϕ	0.84	0.86	0.87	0.88	0.89	0.9	
Eq. Spherical size	1.11	1.02	0.92	0.83	0.76	0.64	
Density ρ (Kg/m^3)	2650	2650	2650	2650	2650	2650	
Surface Area (m^2/g)	2.04E-03	2.23E-03	2.46E-03	2.71E-03	3.00E-03	3.54E-03	
Weight fraction per g (%)	0.74	7.54	8.94	48.63	28.22	5.93	100
Area of each fraction (m^2/g)	1.51E-05	1.68E-04	2.20E-04	1.32E-03	8.46E-04	2.10E-04	**2.78E-03**

APPENDIX 2.2 Contribution of iron(II) adsorption onto iron hydroxide flocs in adsorptive iron removal

Assumptions

1. Influent iron concentration = 5 mg/l
2. Filtration rate = 8 m/h
3. Filter run time = 48 hours
4. 10% of iron(II) is oxidised to form the flocs, before entering the filter bed
   Therefore, the concentration of the iron hydroxide flocs = 0.5 mg iron(III)/l
5. All iron hydroxide flocs formed are retained in the filter bed.
6. The iron(II) concentration and hence the iron(II) adsorption varies along the depth of the filter.
   Average iron(II) adsorption capacity of the lepidocrocite flocs (at pH = 7.0, C_e = 2.5 mg/l)
   = 0.1 mg iron(II) /mg iron(III) (From Fig. 2.7)

Calculations

- Total volume of water filtered per m^2 per run = 8 $m^3/m^2/h \times 48$ h $\times$ 1 m^2 = 384 m^3
- Amount of floc formation in a filter bed
  - a) at t = 0 h, = 0 g
  - b) at t = 48 h, (0.5 $g/m^3 \times 384$ m^3) = 200 g
  - c) average, during a filter run = 100 g
- Total iron(II) adsorption capacity of the flocs formed
  = 100 g [iron(III)] $\times$ 0.1 g iron(II) / g iron(III) = 10 g iron(II).
- Total iron(II) to be removed
  = 5 g /$m^3 \times 384$ m^3 = 2000 g
- The contribution of iron(II) adsorption onto iron hydroxide flocs < 1%

Therefore, the above calculations show that the contribution of iron(II) adsorption onto iron hydroxide flocs formed during a filter run depends on the oxygen present in feed water and consequently the extent of iron(II) oxidation in solution.

Chapter 3

EFFECT OF WATER QUALITY ON IRON(II) ADSORPTION

Part of this Chapter was published by S.K. Sharma, C. Sebwato, B. Petrusevski and J.C. Schippers (2000) *Proceedings of IWA Conference on Innovations in Classical and Conventional Water Treatment Processes,* Amsterdam, The Netherlands. 15, 1-12.

ABSTRACT

Different ions and compounds present in groundwater might influence iron(II) adsorption onto filter media and, hence, affect the process of adsorptive iron removal. Laboratory-scale batch adsorption experiments were conducted to investigate the effect of major inorganic ions present in groundwater namely Mn^{2+}, Ca^{2+}, NH_4^+, SO_4^{2-}, Cl^-, PO_4^{3-}, silica, alkalinity and background ionic strength on the iron(II) adsorption onto new silica sand and iron oxide coated sand. Within the concentration range examined, NH_4^+, Cl^-, alkalinity and background ionic strength had no significant effect on iron(II) adsorption onto both the media tested. An increase in concentration of Mn^{2+}, Ca^{2+} and PO_4^{3-} decreased the iron(II) adsorption onto new sand, whereas an increase in iron(II) adsorption was observed when SO_4^{2-} concentration was increased. The effect of Mn^{2+}, Ca^{2+}, SO_4^{2-} and PO_4^{3-} on iron(II) adsorption onto iron oxide coated sand, however, was minimal. Increasing the concentration of silica from 0 to 40 mg/l decreased iron(II) adsorption onto new sand and iron oxide coated sand by 23% and 16% respectively. In general, the effect of different parameters on iron(II) adsorption was more pronounced on new sand compared to iron oxide coated sand. This indicates that once the iron oxide coating is developed on the surface of the filter media, iron(II) adsorption is not hindered significantly by the presence of other inorganic ions in concentrations common in groundwater.

3.1 INTRODUCTION

Adsorption of iron(II) onto iron hydroxide flocs and iron oxide coated sand (IOCS) is well known (O'Connor 1971; Hult 1973a; Takai 1973; Tamura *et al.* 1976; Braster and Martinell 1988; Appelo *et al.* 1999). Iron oxide coatings develop on the surface of the filter media used in iron removal with time in use. Previous studies showed that iron(II) adsorption onto filter media increases with an increase in pH and, compared to new sand, IOCS has a very high capacity for iron(II) adsorption (see Chapter 2). This high iron(II) adsorption capacity of IOCS can be utilised to improve iron removal in filters by operating the filter predominantly in the adsorptive mode. However, it is unclear how a specific groundwater composition or a specific ion or compound present will influence the effectiveness of adsorptive iron removal.

Several ions present in natural groundwater like Mn^{2+}, Ca^{2+}, NH_4^+, Cl^-, SO_4^{2-}, PO_4^{3-} etc, and other water quality parameters like alkalinity, TOC, and temperature might influence the iron(II) adsorption. The ions present may compete with each other for adsorption sites or may enhance, decrease, or even inhibit the adsorption of iron(II), thereby influencing the overall efficiency of adsorptive iron removal (Weber 1972; Faust and Aly 1998). Secondly, adsorptive iron removal is more effective after the development of an iron oxide coating on filter sand. The above mentioned water quality parameters can also influence the rate of the coating development by affecting adsorption onto the new sand surface. Additionally, ions adsorbed onto the sand surface may influence the physicochemical characteristics of the coating, thus influencing the iron(II) adsorption capacity. These ions might also affect the rate of oxidation of adsorbed iron(II) and consequently the regeneration of adsorption sites. Hence, the knowledge of the effect of water quality parameters on iron(II) adsorption is important for the successful exploitation of the adsorption-oxidation mechanism of iron removal. This study focuses on the effect of some major groundwater quality parameters, namely Mn^{2+}, Ca^{2+}, NH_4^+, Cl^-, SO_4^{2-}, PO_4^{3-}, silica, alkalinity, and background ionic strength on the adsorption of iron(II) onto new silica sand and IOCS.

3.2 THEORETICAL BACKGROUND

3.2.1 Adsorption of ions on hydrated oxide surface

The hydrated surface of oxides (silica sand and IOCS) can adsorb different cations and anions present in the water. The extent of adsorption or surface complexation depends on the type and density of the adsorption sites available and the nature of the adsorbing ion. As Fig. 3.1 illustrates, a cation can associate with a surface as an inner-sphere or outer-sphere complex, depending on whether a chemical bond between the metal and the electron-donating oxygen ions is formed or if a cation of opposite charge approaches the surface groups to a critical distance. In the case of an outer-sphere complex, as with the solute ion pairs, the cation and the base are separated by one or more water molecules. Furthermore, ions may be in the diffuse swarm of the electric double layer.

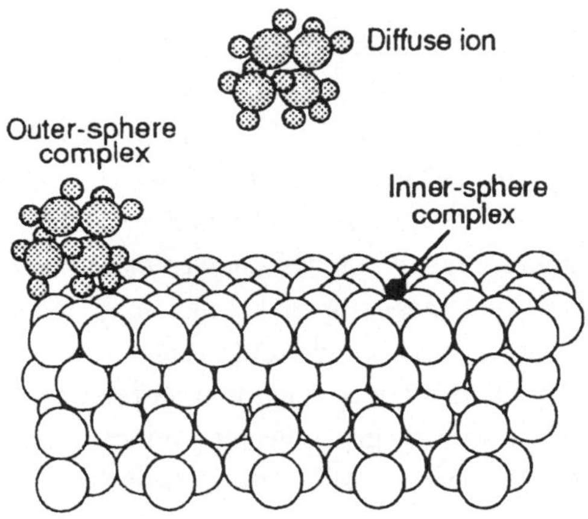

Fig. 3.1 Surface complex formation of an ion on the hydrous oxide surface. The ion may form an inner-sphere complex ("chemical bond"), an outer-sphere complex ("ion pair") or be in diffused swarm of the electrical double layer (Stumm and Morgan 1981).

A simple method of distinguishing between inner-sphere and outer-sphere complexes is to assess the effect of ionic strength on the surface complex formation equilibria. A strong dependence of ionic strength is typical for an outer-sphere complex. Furthermore, outer-sphere complexes involve electrostatic bonding mechanisms and therefore are less stable than inner surface complexes, which necessarily involve appreciable covalent bonding along with ionic bonding.

Cation surface complexation by hydrous oxides involves formation of bonds with surface oxygen atoms and the release of protons from the surface, as represented by equations (3.1) and (3.2)

$$\equiv S\text{-}OH + M^{2+} \leftrightarrows \equiv S\text{-}OM^{+} + H^{+} \quad (3.1)$$

or equivalently

$$\equiv S\text{-}OH + M^{2+} + H_2O \leftrightarrows \equiv S\text{-}OMOH_2^{+} + H^{+} \quad (3.2)$$

where $\equiv$S-OH and M^{2+} represent a hydrated oxide surface and a divalent cation respectively (Dzombak and Morel 1990).

Anion adsorption by hydrous oxide occurs via ligand exchange reactions in which hydroxyl surface groups are replaced by the sorbing ions (Dzombak and Morel 1990)

$$\equiv S\text{-}OH + A^{2-} + H^{+} \leftrightarrows \equiv S\text{-}A^{-} + H_2O \quad (3.3)$$

and/or

$$\equiv S\text{-}OH + A^{2-} + 2H^{+} \leftrightarrows \equiv S\text{-}HA + H_2O \quad (3.4)$$

where A^{2-} represents a divalent anion.

Ionic charge and size play an important role in cation sorption. Other things being equal, the adsorption affinity is greater for the ion with higher valency. For the ions of the same valence the affinity for adsorption increases with atomic number and apparently decreases with decreasing ionic size (Matthess 1982). Simultaneous adsorption of different cations involves competition between the cations for the available adsorption sites, since adsorption from solution is usually restricted to a monolayer (Yang 1997; Faust and Aly 1998). In the presence of other cations commonly present in groundwater like Ca^{2+}, Mn^{2+} etc, the adsorption sites available for iron(II) may be reduced and therefore a reduction in its adsorption is expected. The degree of mutual inhibition is related to relative sizes, concentrations, and adsorption affinities of the competing molecules (Weber 1972, 1985).

There are four ways in which ligands, which are mostly anions, can affect adsorption of metal ions onto oxide surfaces (Benjamin and Leckie 1982):

i) Metal-ligand complexes may form in solution and adsorb only weakly or not at all. In this case, formation of soluble complexes may be considered to compete with reactions forming "surface complexes", and adsorption is decreased compared to the ligand-free system.

ii) Metal ion and ligand may interact indirectly at the surface, thereby altering the surface electrical properties. For instance, if the ligand is an anion, it may adsorb and make the coulombic interaction between the solid and a cationic metal ion more favourable.

iii) The metal-ligand complex may adsorb strongly, thereby enhancing the removal of metal, ligand, or both from the solution compared to the case where either one is present alone.

iv) The complex formed may have no effect on metal adsorption. In this case, the interaction of the surface with the metal-ligand complex is the same as that with the free uncomplexed metal.

Hence, depending on the nature of the anions present and complexes formed, iron(II) adsorption onto filter media may be enhanced, decreased, or unaltered. Table 3.1 summarises the effect of various parameters on adsorption in systems with multiple adsorbates.

Table 3.1 Effect of various parameters on sorption in systems with multiple adsorbates

Parameter	Effect of increasing parameter on metal (M) sorption
PH	Typically a dramatic increase (decrease) in sorption of cations (anions) as pH increases through a critical pH range
Any other adsorbate (Ad)	Competition for adsorption sites; M sorption decreases
Any ionized adsorbate (Ad^{m+} or Ad^{p-})	Alteration of potential in electrical double layer; effect depends on overall stiochiometry of sorption reactions; normally sorption of cations makes surface more attractive to anions and less attractive to other cations
Sparingly soluble adsorbate	New surface phase may be formed; surface phase may consist of adsorbate plus M^{n+}, OH^-, constituent ion of adsorbent, or some combination of these; effect on M depends on adsorbent properties of new phase
Adsorbates which can form a soluble complex with metal	Depends on relative adsorptive bond strength of complexes as compared to uncomplexed ions

Source: Benjamin (1983)

3.2.2 Ionic strength and iron(II) adsorption

The ionic strength (I) describes the intensity of the electric field in the solution and is mathematically calculated by the equation

$$I = \frac{1}{2}\sum_i c_i z_i^2 \tag{3.5}$$

where c_i is the concentration of the ionic species i (moles/litre) and z_i is the charge of the species i (Snoeyink and Jenkins 1980). The units of ionic strength I is therefore also moles/litre.

The adsorption of a solute on an adsorbent is a result of two forces; the chemical force and the coulombic force. The chemical forces lead to covalent bonds, hydrophobic bonds, hydrogen bridges and steric or orientation effects, while the coulombic forces lead to electrostatic and van der Waals attractions (Stumm and Morgan 1981). The surface charge is dependent on ionic strength. The coulombic forces are due to ionic interactions and therefore depend on the ionic strength. This implies that ionic strength of the water may influence the adsorption of iron(II) onto the surface of the filter media.

3.3 MATERIALS AND METHODS

3.3.1 Filter media

New (virgin) silica sand and IOCS developed in the laboratory were used in this study. The characteristics of the filter media used are presented in Table 3.2. The new sand was washed in demineralised water, soaked in hydrochloric acid (pH < 2) for 24 hours, again rinsed with demineralised water and finally dried at 40^0C.

The IOCS was developed by running a pilot filter at a pH of 7.1 ± 0.1 with feed water iron(II) concentration of 6.0 ± 0.2 mg/l at a filtration rate of 5 ± 0.5 m/h over a run time of 1500 hours. The details of the process of coating development and characteristics of the coating are presented elsewhere (see Chapter 5). The IOCS developed was washed gently with demineralised water to remove any deposits and dried at room temperature before using for batch experiments.

Table 3.2 Characteristics of the new sand and IOCS used

Media	New sand	IOCS
Grain size (mm)	0.8-1.25	0.8-1.25
Effective size d_{10} (mm)	0.82	0.82
Specific surface area (BET) (m^2/g)	1.0	7.7
Surface extractable iron content (mg/g sand)	0	22.3

3.3.2 Experimental set-up

Iron(II) adsorption onto new sand and IOCS was measured at different concentrations of the parameters under study by conducting batch adsorption equilibrium experiments at room temperature (18 - 20° C). The experimental set-up and the procedure used to determine iron(II) adsorption onto filter media was similar to that used in an earlier study (see Chapter 2). Nitrogen gas was used for deoxygenation. The pH was adjusted using HCO_3^--CO_2 buffer. The pH and oxygen concentration were monitored continuously with a pH meter (WTW pH 323) and oximeter (WTW Oxi 340) to ensure constant pH and anoxic conditions.

Table 3.3 shows the different chemicals used to introduce different ions to be studied in the model groundwater. All the chemicals used were of reagent grade.

Table 3.3 Chemicals used to introduce different ions and their concentration range

Chemical	Purpose	Concentration range
$NaHCO_3$	Alkalinity and pH control	1 - 8 mmol HCO_3^-/l
$FeSO_4.7H_2O$	Iron(II) source	4 mg Fe^{2+}/l
$MnSO_4.H_2O$	Manganese(II) source	0.25 - 1.5 mg Mn^{2+}/l
$Ca(NO_3)_2.4H_2O$	Calcium source	50 - 200 mg Ca^{2+}/l
NH_4Cl	Ammonium source	0.25 - 1.0 mg NH_4^+/l
Na_2SO_4	Sulphate source	6 - 200 mg SO_4^{2-}/l
NaCl	Chloride source	10 - 80 mg Cl^-/l
$Na_3PO_4.12H_2O$	Phosphate source	0.25 - 1.0 mg PO_4^{3-}/l
$NaSiO_3.5H_2O$	Silica source	10 - 40 mg SiO_2/l
$NaNO_3$	Background ionic strength	2.5 - 25 mmol $NaNO_3$/l

3.3.3 Experimental procedure

Initially 1.5 litres of demineralised water was deoxygenated by bubbling nitrogen gas through the solution. The pH was adjusted to the desired level with HCO_3^- and CO_2 gas. Then 15 ml of 400 mg/l deoxygenated iron(II) stock solution and different amounts of stock solution of the parameter under study were dosed into the closed reactor containing deoxygenated water and 40-50 g of filter media. The pH during the experiment was controlled to within 0.05 units. On equilibrium, concentrations of iron(II) and the parameter under study in the reactor were measured (Standard Methods 1995) and the amount of iron(II) adsorbed onto the media was calculated by a mass balance. All the experiments were carried out in duplicate to verify the reproducibility of the results. For comparison of the effect of a given water quality parameter on new sand and IOCS, iron(II) adsorbed in the presence of different concentrations of the parameter under study were expressed as the relative percentage of iron(II) adsorption onto given media in the absence of that parameter.

The equilibration time for each filter media was determined separately by measuring the hourly variation in iron concentration in the solution. Equilibrium was considered to have been reached when the difference in iron concentrations of two consecutive hourly samples was less than 0.05 mg/l. Under the given experimental conditions, the equilibration times for new sand and IOCS were 4 hours and 7 hours respectively.

3.4 RESULTS AND DISCUSSION

3.4.1 Effect of manganese

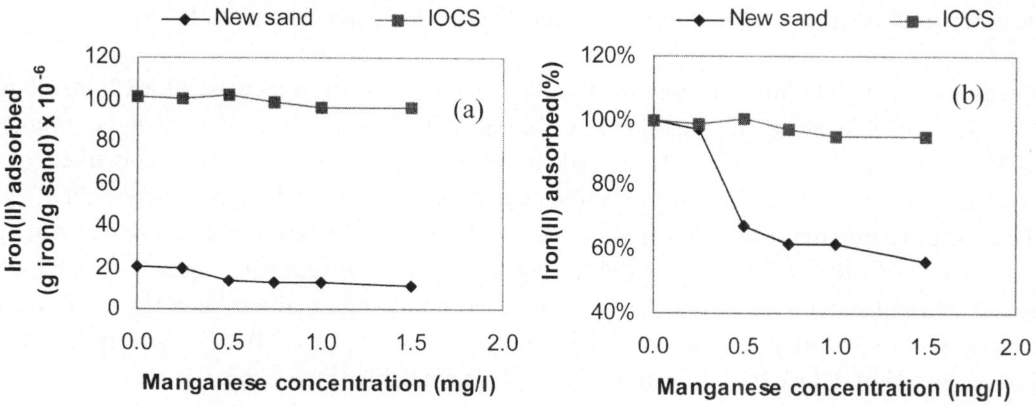

Fig. 3.2 Effect of Mn^{2+} concentrations on iron(II) adsorption onto new sand and IOCS (pH = 6.8 ± 0.05, initial iron concentration = 4.0 mg/l)

Figure 3.2 presents the influence of Mn^{2+} concentration on iron(II) adsorption onto filter media. For comparison of the effects on new sand and IOCS, the iron(II) adsorption at different manganese concentrations were expressed as the relative percentage of iron(II) adsorption onto given media in the absence of manganese (Fig 3.2b). Iron(II) adsorption decreased by 44% for new sand and by 6% for IOCS when the Mn^{2+} concentration increased from 0 mg/l to 1.5 mg/l. This reduction in iron(II) adsorption in the presence of Mn^{2+} may be due to competitive adsorption between iron and manganese ions. Additionally, the introduction of positively charged ions reduces the negative charge of the sand surface resulting in the reduced coulombic force and hence reduced iron(II) adsorption.

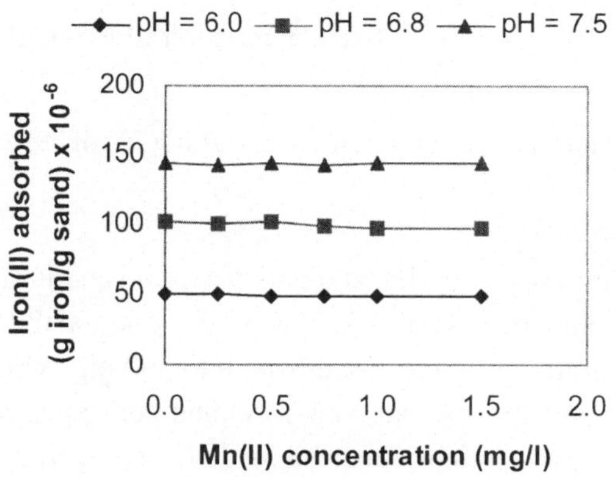

Fig. 3.3a Effect of Mn(II) concentration on iron(II) adsorption onto IOCS at different pH (initial iron concentration = 4.0 mg/l)

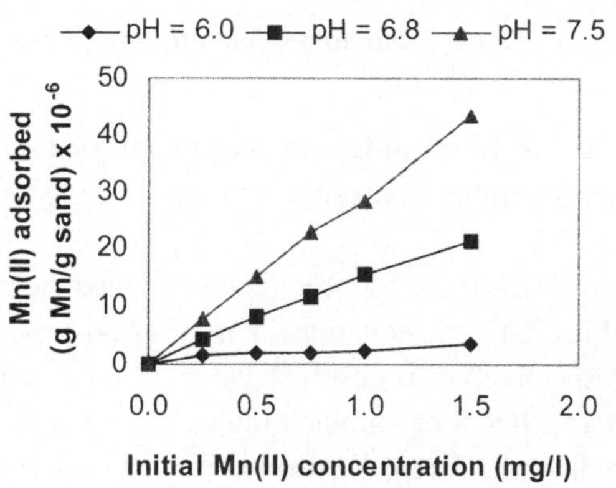

Fig. 3.3b Adsorption of Mn(II) onto IOCS at different pH (initial iron concentration = 4.0 mg/l)

Figure 3.3a presents the effect of Mn(II) concentration on iron(II) adsorption onto IOCS at different pH. At all three pH studied, there was no measurable effect of increasing the Mn(II) concentration on iron(II) adsorption. Monitoring of the Mn^{2+} concentration in the solution confirmed that Mn^{2+} is adsorbed onto IOCS (Fig. 3.3b), however there was no measurable adsorption of Mn^{2+} onto new sand. Adsorption of Mn^{2+} onto IOCS but not on new sand has also been reported in other studies (Graveland and Heertjes 1975; Knocke *et al.* 1988).

It was also found that Mn^{2+} adsorption onto IOCS increases with an increase of pH from 6.0 to 7.5. This is in agreement with the findings of Morgan (1964) and Weber (1972) that the sorption of Mn^{2+} onto iron hydroxides is pH-dependent, increasing as the pH increases and markedly so after the pH reaches 8.2. These results indicate that the presence of manganese does not affect the adsorptive iron removal in filters with IOCS and, additionally, some manganese removal by adsorption onto IOCS can be achieved during adsorptive iron removal. Because of the high adsorption capacity of IOCS and the low Mn^{2+} concentration normally present in groundwater (manganese to iron in groundwater normally ranges from 1/10 to 1/20), Mn^{2+} adsorption onto IOCS would not reduce the adsorption of iron(II) appreciably (Fig. 3.2b).

3.4.2 Effect of calcium

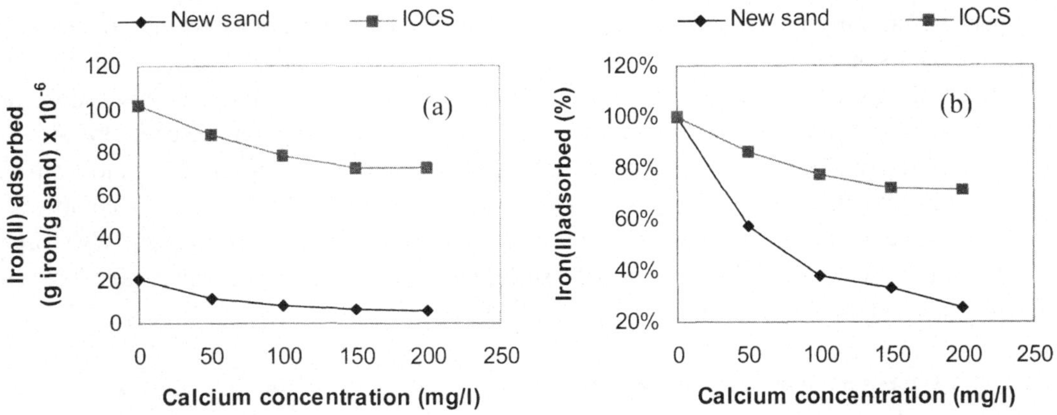

Fig. 3.4 Effect of Ca^{2+} concentrations on iron(II) adsorption onto new sand and IOCS (pH = 6.8 ± 0.05, initial iron concentration = 4.0 mg/l)

An elevated Ca^{2+} concentration inhibited the adsorption of iron(II) onto both new sand and IOCS (Fig. 3.4). A reduction in iron(II) adsorption onto new sand and IOCS of 74% and 29% respectively was observed when the Ca^{2+} concentration was increased from 0 to 200 mg/l (Fig. 3.4b). IOCS has a much higher iron(II) adsorption capacity than new sand and consequently reduction of iron(II) adsorption due to calcium ions was less pronounced for IOCS compared to new sand.

Ca^{2+} and Fe^{2+} are both divalent ions, therefore their competition for the adsorption sites may depend on their sizes and concentrations. The calcium ion is relatively large (ionic radius of Ca^{2+} is 0.099 nm compared to 0.077 nm of Fe^{2+}), can be hydrated, and forms various aqueous complexes with other inorganic ions like HCO_3^-, OH^-, CO_3^{2-}, SO_4^{2-}, PO_4^{3-} (Matthess 1982; Hem 1989). In addition, the Ca^{2+} concentration was much higher than the iron(II) concentration in all cases and hence iron(II) adsorption inhibition could be expected. At a pH of 6.8 and alkalinity of 6 mmol HCO_3^-/l for all calcium concentrations studied, the Langlier saturation index (SI) was negative, indicating that no precipitation of $CaCO_3$ onto filter media could be expected. No measurable change in calcium concentration was observed during the experiment confirming previous assumptions.

Experimental results suggest that adsorptive iron removal from hard groundwaters may be rather less efficient as the calcium ion has a negative effect on iron(II) adsorption. This is more likely if calcium has no effect on the rate of oxidation of adsorbed iron(II). However, for waters with a calcium concentration of 50 mg/l, adsorption of iron(II) onto IOCS was only reduced by about 10%. In comparison, the performance of floc filtration iron removal from soft waters with low alkalinity is poor (Hult 1973a,b) due to slow iron(II) oxygenation and formation of weak flocs. In such cases, adsorptive iron removal may be a better option. Conversely, floc filtration iron removal is better and easier in hard waters with substantial alkalinity than in soft waters because of the possible siderite ($FeCO_3$) precipitation and increased oxygenation rate (Hult 1973b, Cleasby 1975).

3.4.3 Effect of ammonium

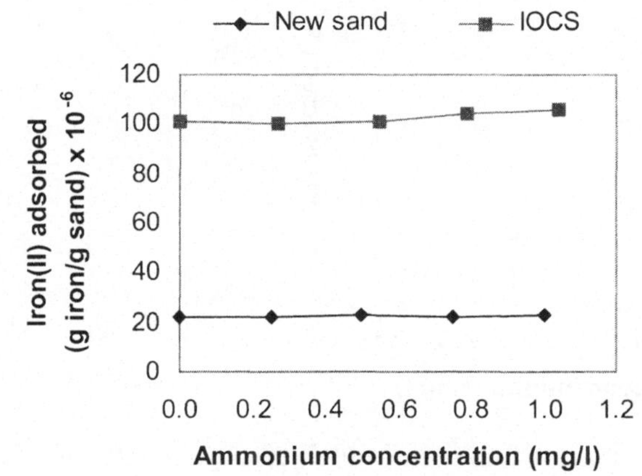

Fig. 3.5 Effect of NH_4^+ concentrations on iron(II) adsorption onto new sand and IOCS (pH = 6.8 ± 0.05, initial iron concentration = 4.0 mg/l)

An increase in ammonium concentration from 0 mg/l to 1.0 mg/l had no significant effect on the adsorption of iron(II) onto both new sand and IOCS (Fig. 3.5). Furthermore, no adsorption of ammonium onto new sand was detected. The new sand surface preferentially adsorbs divalent ions like iron(II) as opposed to monovalent ions like ammonium. Moreover, new sand has a low iron(II) adsorption capacity compared to IOCS (see Chapter 2). Ammonium does not form complexes with iron(II) and hence it does not influence iron(II) speciation. Therefore, the presence of ammonium has no effect on the adsorption of iron(II) onto new sand. It was also observed that some ammonium was adsorbed onto IOCS when the ammonium concentration exceeded 0.25 mg/l. The ammonium adsorption capacity of IOCS under batch conditions tested was found to be 7.0 x 10^{-6} g ammonium/g IOCS at an initial ammonium concentration of 1 mg/l.

3.4.4 Effect of sulphate

Figure 3.6 illustrates the influence of SO_4^{2-} on iron(II) adsorption onto new sand and IOCS. In contrast to divalent cations, SO_4^{2-} at lower concentrations (up to 50 mg/l) increased the iron(II) adsorption capacity of both new sand and IOCS. At a SO_4^{2-} concentration of 50 mg/l, the iron(II) adsorption increased by 47% and 5% for new sand and IOCS respectively (Fig. 3.6b). However, a further increase of SO_4^{2-} concentrations above 50 mg/l resulted in a reduction of iron(II) adsorption onto both new sand and IOCS, probably due to complexation of iron(II) with SO_4^{2-}. Nevertheless, even at the highest SO_4^{2-} concentration tested (207 mg/l), iron(II) adsorbed was 23% more on the new sand and only 4% less on IOCS in comparison to that at SO_4^{2-} concentration of 7 mg/l (Fig. 3.6b). Hence, there is a considerable increase in iron(II) adsorption onto new sand in the presence of SO_4^{2-}, and no significant SO_4^{2-}-induced hindrance on the iron(II) adsorption onto IOCS.

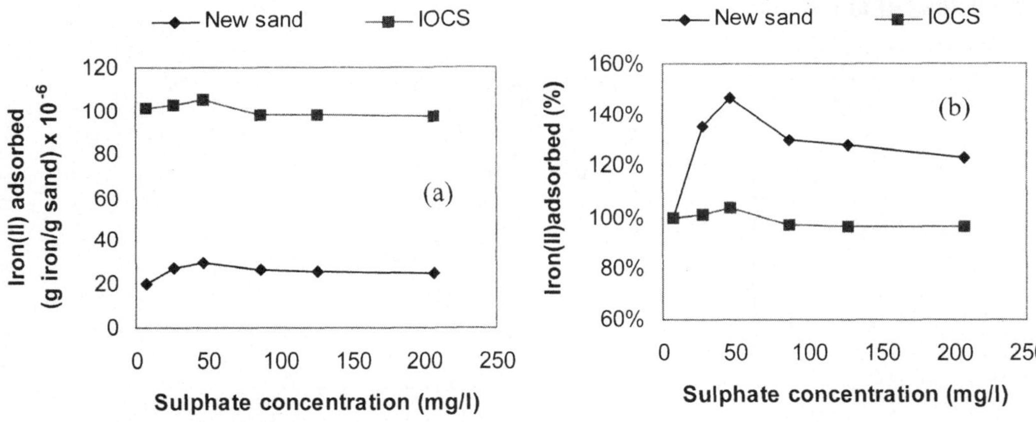

Fig. 3.6 Effect of SO_4^{2-} concentrations on iron(II) adsorption onto new sand and IOCS (pH = 6.8 ± 0.05, initial iron concentration = 4.0 mg/l)

The initial increase in iron(II) adsorption with an increase in sulphate concentration may be due to sulphate adsorption onto the sand oxide surface. Sulphate adsorption onto filter media changes the oxide surface electrical properties (making it more negatively charged), thus promoting coulombic interaction between the media surface and iron(II).

The sulphate ions has a tendency to form an "ion pair" (a special type of inter-ionic association involving two ions of opposite charge). As sulphate concentrations increase, an increasing proportion of the sulphate in solution becomes tied up in this way (Hem 1989). The eventual decrease in iron(II) adsorption at a sulphate concentration >50 mg/l may be due to formation of iron(II)-sulphate ion pairs, i.e., sulphate ions start competing with the oxide surface for the available iron(II) ions. This increases the solution activity of the ion pair, but reduces the amount of iron(II) available for adsorption. Benjamin and Leckie (1982) also reported a reduction in cadmium adsorption onto iron oxide with an increase in sulphate concentration.

The increased iron(II) adsorption onto new sand at high sulphate concentrations is an indication of the importance of coulombic interaction in the case of new sand. In the case of IOCS, only a 5% increase was observed, re-emphasising the negligible contribution by the coulombic forces in the adsorption of iron(II) onto IOCS. The point of zero charge of hydrous iron oxide lies between 7.8 and 8.2 (Dzombak and Morel 1990). It should be noted that at a pH of 6.8, IOCS has a positive surface charge implying that adsorption of iron(II) is taking place against coulombic forces.

3.4.5 Effect of chloride

For both new sand and IOCS, no significant effect on iron(II) adsorption was observed when the chloride concentration was increased from 0 to 80 mg/l (Fig. 3.7). Additionally, no measurable decrease in the chloride concentration was observed during the experiments. Chloride ions being monovalent are generally located in the diffuse layer and the bulk solution (Stumm and Morgan 1981) and may not form inner-sphere bonds with hydrous oxide surfaces. Therefore, they do not influence iron(II) adsorption, which is adsorbed to specific oxide surface sites.

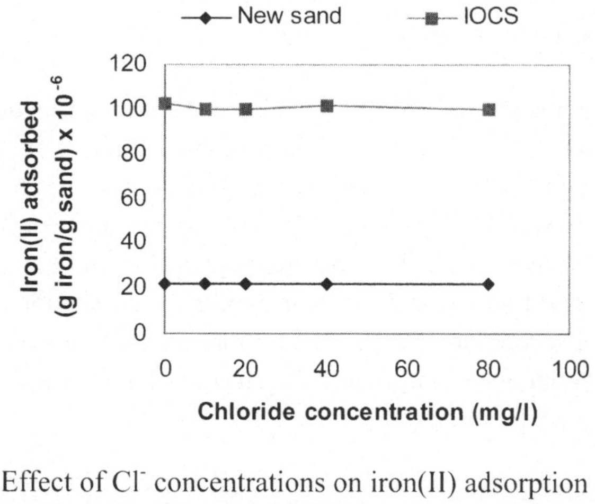

Fig. 3.7 Effect of Cl^- concentrations on iron(II) adsorption onto new sand and IOCS (pH = 6.8 ± 0.05, initial iron concentration = 4.0 mg/l)

Hem (1989) mentioned that chloride can form ion pairs and complex ions with some of the cations present in the natural waters, but these complexes are not strong enough to be of significance in the chemistry of freshwaters. The lack of influence of chloride on iron(II) adsorption may be attributed to the iron(II)-chloride complexes formed, if any, being adsorbed onto filter media similarly as uncomplexed iron(II) ions.

3.4.6 Effect of phosphate

Figure 3.8 presents the influence of PO_4^{3-} concentration on iron(II) adsorption onto filter media. A 15% decrease in iron(II) adsorption onto new sand was observed at the minimum tested phosphate concentration of 0.25 mg/l. At a phosphate concentration of 1.0 mg/l, the iron(II) adsorption was reduced by 23%. However, in the case of IOCS there was no significant change in iron(II) adsorption with the presence of phosphate (0.25 mg/l - 1.0 mg/l).

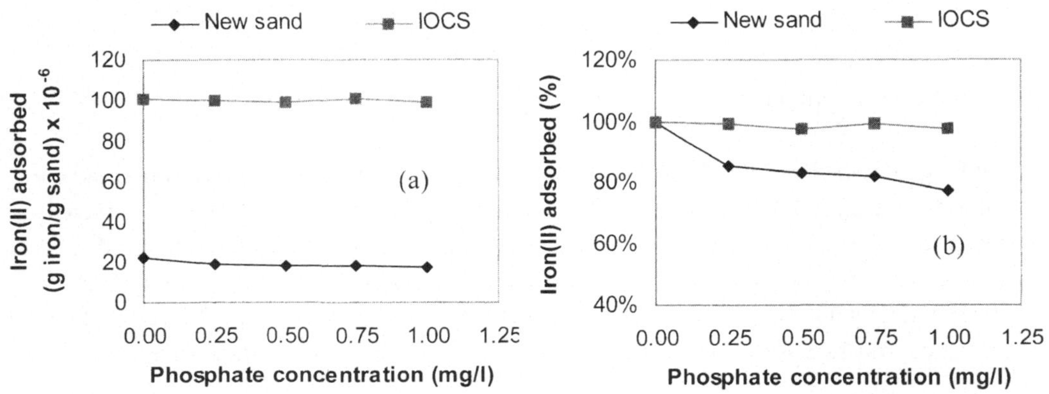

Fig. 3.8 Effect of PO_4^{3-} concentrations on iron(II) adsorption onto new sand and IOCS (pH = 6.8 ± 0.05, initial iron concentration = 4.0 mg/l)

It was observed that some phosphate was also adsorbed onto filter media along with the iron(II). In the case of new sand, phosphate adsorption ranged from about 19% at 0.25 mg/l phosphate to about 7% at 1 mg/l phosphate. At phosphate concentrations of 0.25 mg/l and 1 mg/l, phosphate adsorption onto IOCS was 100% and 96% respectively. Phosphate adsorption onto iron oxides and iron oxide coated sand has also been reported by others (Broberg 1984; Vaithiyanathan and Corell 1992; Berner and Rao 1994). Despite the complete adsorption of phosphate onto IOCS, no influence was observed on iron(II) adsorption. This is probably because iron(II) and phosphate are ions of opposite charges and they have different specific adsorption sites on the surface of the filter media.

Phosphate can bind specifically to metal oxides and alter the surface charges of oxides at high concentrations (Stumm and Morgan 1996). Furthermore, specifically adsorbed anions decrease the surface charge and potential of the oxide surface making it less favourable for anion adsorption, but promoting cation adsorption (Dzombak and Morel 1990). Phosphate adsorption onto filter media enhances the coulombic interaction, therefore an increase in iron(II) adsorption could be expected. No significant effect of phosphate on iron(II) adsorption onto IOCS is likely due to the low concentration of phosphate used in this study such that the chemical forces are more significant than the coulombic forces.

3.4.7 Effect of silica

Increasing the concentration of silica in model groundwater decreased iron(II) adsorption onto both new sand and IOCS (Fig. 3.9). At concentrations of silica examined in this study (10 mg/l-40 mg/l), the iron(II) adsorption onto new sand was reduced by 14%-27%. In the case of IOCS, iron(II) adsorption reduced by 16% at a silica concentration of 10 mg/l. A further increase in the silica concentration from 10 mg/l to 40 mg/l resulted in no significant decrease (<1%) in iron(II) adsorption. It was found that some silica is also adsorbed onto filter media together with iron(II).

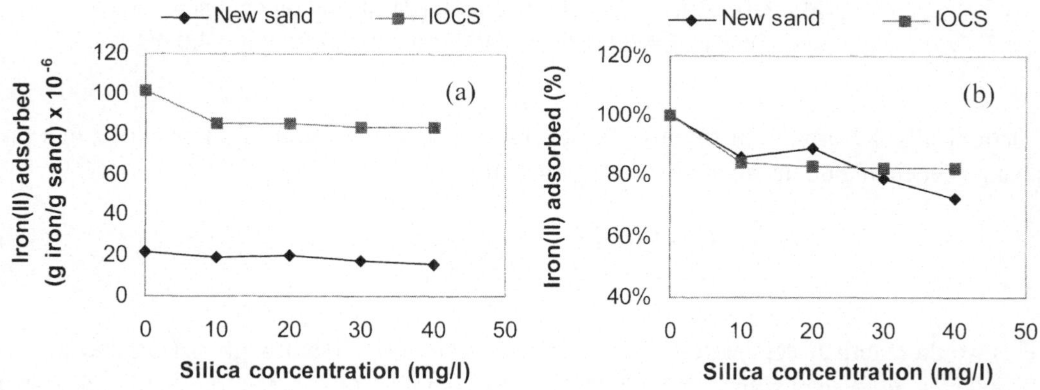

Fig. 3.9 Effect of silica concentrations on iron(II) adsorption onto new sand and IOCS (pH = 6.8 ± 0.05, initial iron concentration = 4.0 mg/l)

The decrease in iron(II) adsorption in the presence of silica was likely due to i) the adsorption of silica onto filter media and/or ii) the formation of iron(II)-silica complexes. The adsorption of silica changes the structure and surface chemistry of iron oxides and lowers the surface charge (Schwertmann and Fetcher 1982; Anderson and Benjamin 1985; Meng and Letterman 1993). The adsorption of the silica, therefore, might have reduced some adsorption sites freely available for the iron(II). Secondly, the formation of iron(II)-silica complexes reduces the freely available iron(II) ions in the solution (Schenk and Weber 1968). It is likely that the iron(II)-silica complex is relatively less adsorbed onto filter media than free or uncomplexed iron(II) ion.

3.4.8 Effect of alkalinity

An increase in alkalinity from 1 mmol HCO_3^-/l to 8 mmol HCO_3^-/l did not show any significant effect on the adsorption of iron(II) onto filter media (Fig. 3.10). It is to be noted that in these experiments the pH was kept constant.

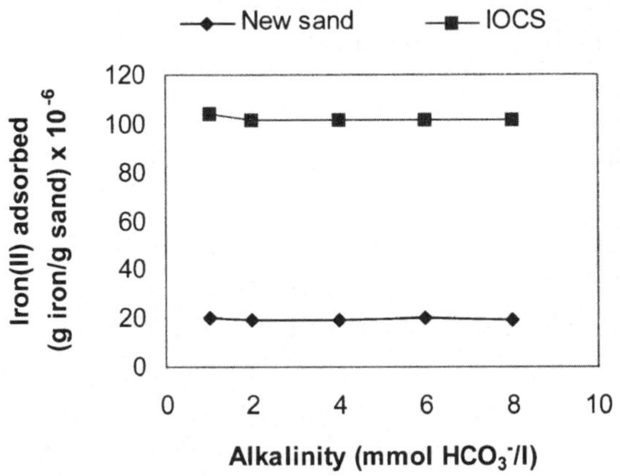

Fig. 3.10 Effect of alkalinity on iron(II) adsorption onto new sand and IOCS (pH = 6.8 ± 0.05, initial iron concentration = 4.0 mg/l)

Theoretically, adsorption of cations liberates protons as per equation (3.1), resulting in a pH change depending on the alkalinity (buffer capacity).

$$H^+ + HCO_3^- \leftrightarrows H_2O + CO_2 \quad (3.6)$$

$$H^+ + CO_3^- \leftrightarrows HCO_3^- \quad (3.7)$$

This would in turn affect the iron(II) adsorption capacity, which is strongly pH-dependent (see Chapter 2). For waters with a high alkalinity of 5.0 mmol HCO_3^-/l, under the conditions tested the resulting pH change with iron(II) adsorption would be 0.02 unit for IOCS and negligible for new sand. In the case of waters with a low alkalinity of 1.0 mmol HCO_3^-/l, the pH changes would be 0.08 unit for IOCS and 0.02 unit for new sand, which are still insignificant.

Iron removal by the floc formation-filtration mechanism gives poor results at low alkalinity due to slow oxidation and poor floc formation (Robinson and Breland 1968; Hult 1973a,b; Olson and Twardowski 1975; Cleasby 1975). Secondly, iron content in groundwater is inversely proportional to the alkalinity due to siderite ($FeCO_3$) solubility limitations (O'Connor 1971) and a higher iron content is expected in groundwater with low alkalinity. The experimental results demonstrated that alkalinity has no significant effect on iron(II) adsorption onto either new sand or IOCS. This implies that, unlike floc filtration iron removal, adsorptive iron removal could be a viable option even for groundwater with low alkalinity.

3.4.9 Effect of background ionic strength

While studying the effect of a certain parameter, ions other than those targeted were also introduced in the model water. In order to ensure that the observed effects were actually due to particular parameters and not due to the change in the ionic strength of water, the effect of ionic strength on iron(II) adsorption onto filter media was examined. All the anions studied in the experiments were introduced as sodium salts. Nitrate is known as a fairly inert and non-complexing ligand. Hence, $NaNO_3$ was selected to vary the background ionic strength of the model groundwater.

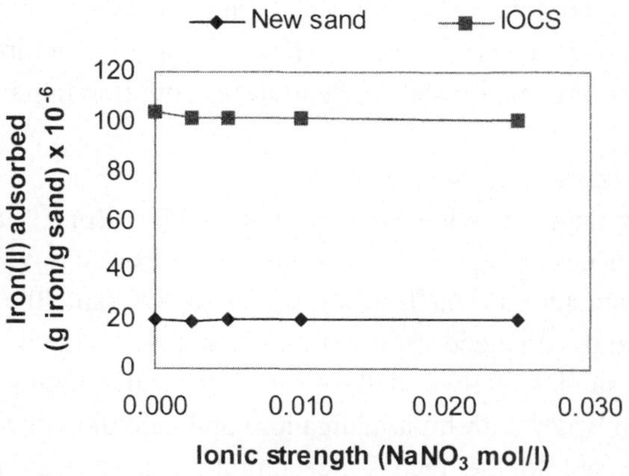

Fig. 3.11 Effect of background ionic strength on iron(II) adsorption onto new sand and IOCS (pH = 6.8 ± 0.05, initial iron concentration = 4.0 mg/l)

It is clear from Fig. 3.11 that an increase in background ionic strength from 0 to 0.025 M $NaNO_3$ had no effect on iron(II) adsorption onto new sand and IOCS. This is consistent with the findings of other researchers that ionic strength has no effect on cation adsorption onto oxide surfaces (Swallow *et al.* 1980; Hayes and Leckie 1987; Dzombak and Morel 1990). These results also imply that neither Na^+ nor NO_3^- ions had any influence on iron(II) adsorption onto filter media.

Theoretically, if the coulombic component of the adsorption force were of any importance, one would expect ionic strength to influence the adsorption of iron(II) onto filter media. However, the range of ionic strength considered in this study is relatively low (up to 0.025 M) but typical for natural groundwaters. Dzombak and Morel (1990) mentioned that the absence of the effect of ionic strength on cation adsorption is because the cation adsorption is mainly governed by the chemical forces and occurs against the electrostatic or coulombic forces.

3.4.10 Practical implications

Experimental results revealed that ammonium, chloride, alkalinity and background ionic strength have no effect on adsorptive iron removal. It was also found that within the concentration range of natural groundwater, the effect of manganese, calcium, sulphate and phosphate on iron(II) adsorption onto IOCS is minimal. For groundwaters containing both manganese and sulphate at concentrations of less than 1.0 and 50 mg/l respectively, the net effect on iron(II) adsorption onto new sand is likely to be insignificant as the decrease in iron(II) adsorption due to manganese is compensated by an increase in iron(II) adsorption due to sulphate. It was also observed that some manganese and ammonium removal takes place during adsorptive iron removal. This implies that in adsorptive iron removal, adsorption of iron(II) onto IOCS is not hindered significantly due to the presence of other inorganic ions at concentrations common in natural groundwater.

Development of iron oxide coating on filter sand

The development of an iron oxide coating with a high iron(II) adsorption capacity is a prerequisite for the successful application of adsorptive iron removal. With knowledge of the influence of each parameter on iron(II) adsorption onto new sand, the coating development can be pre-determined and enhanced if required. The results obtained suggest that coating development on new sand is likely to be faster in water with a high sulphate concentration and presumably slower in water with high manganese and calcium concentrations. Therefore, the addition of sulphate salt to the feed water might be a strategy to speed up the coating development on new sand. However, it should be noted that this is likely only when different ions present in groundwater have no significant effect on the rate of oxidation of adsorbed iron(II).

The presence of other groundwater constituents not considered in this study might also affect the adsorption process. Natural organic matters, which have a high complexing ability, may significantly hinder or enhance iron(II) adsorption onto filter media depending upon the relative concentrations of iron(II) and organic matter, and the characteristics of the organic matter (Tipping 1981; Tipping and Cooke 1982; Gu *et al.* 1994, 1995). Some preliminary experiments were conducted using a commercial humic acid (distributed by ACROS B.V., Belgium) to study the effect of organic matter on iron(II) adsorption onto filter media. Increasing the organic matter concentration from 0 mg TOC/l to 1.5 mg TOC/l decreased iron(II) adsorption onto new sand and IOCS by 20% and 30% respectively. Further increase in organic matter from 1.5 mg TOC/l to 6.0 mg TOC/l resulted in no further decrease in iron(II) adsorption onto both new sand and IOCS. It was observed that some organic matter was also absorbed onto IOCS. However, it should be noted that the nature of these organic compounds and their effect is most likely site specific and might be much different from the commercial humic acid. So further detailed investigation is needed to analyse the potential role of organic matter in adsorptive iron removal.

Secondly, the rate of oxidation of iron(II) adsorbed onto filter media might be influenced by the presence of other ions in water. Additionally, the adsorption of "foreign species" on iron oxide coating may impair the adsorption capacity of the coating in the long term. An understanding of the rate and factors affecting the regeneration of iron(II) adsorption sites is, therefore, equally important which needs a detailed study.

3.4 CONCLUSIONS

The effect of Mn^{2+}, Ca^{2+}, NH_4^+, Cl^-, SO_4^{2-}, PO_4^{3-}, silica, alkalinity, and background ionic strength on the adsorption of iron(II) onto new silica sand and IOCS was investigated by conducting batch adsorption experiments. Based on the experimental results, the following conclusions can be drawn:

- In the concentration range examined, NH_4^+ (0.25-1 mg/l), Cl^- (10-80 mg/l), alkalinity (1-8 mmol HCO_3^-/l), and background ionic strength (2.5-25 mmol $NaNO_3$/l), had no significant effect on iron(II) adsorption onto either new silica sand or iron oxide coated sand.
- Increasing the Mn^{2+} concentration from 0 to 1.5 mg/l decreased iron(II) adsorption onto new sand and IOCS by 44% and 6% respectively.
- Iron(II) adsorption onto new sand and IOCS reduced by 74% and 29% respectively when Ca^{2+} concentration was increased from 0 to 200 mg/l.
- An increase in iron(II) adsorption onto new sand was observed when SO_4^{2-} concentration was increased, however the effect of SO_4^{2-} on iron(II) adsorption onto IOCS was minimal.
- Increasing PO_4^{3-} concentration from 0 to 1 mg/l decreased iron(II) adsorption onto new sand by 23%, however there was no significant effect on iron(II) adsorption onto IOCS.
- Iron(II) adsorption onto new sand and IOCS reduced by 27% and 16% respectively when silica concentration was increased from 0 to 40 mg SiO_2/l.
- In general, it was observed that the effect of different ions and compounds on iron(II) adsorption is more pronounced on new silica sand than on IOCS. Once the iron oxide coating is developed on the filter media, the adsorption of iron(II) is not hindered significantly by the presence of other inorganic ions within the concentration range common for groundwater.
- Preliminary experiments with commercial humic acids showed a decrease in iron(II) adsorption capacity of both new and iron oxide coated sand with increasing concentration of organic matter. The effect of organic matters on the iron(II) adsorption capacity needs to be measured in groundwater from different locations since the nature of these organic compounds and their effect is most likely site specific.

REFERENCES

Anderson, P. and Benjamin, M. (1985) Effect of silicon on crystallisation and adsorption properties of ferric oxides. *Environmental Science and Technology*, **19** (11), 1048-1053.

Appelo, C.A.J., Drijver, B., Hekkenberg, R. and de Jonge, M. (1999) Modeling in situ iron removal from ground water. *Ground Water*, **37** (6), 811- 817.

Benjamin, M.M. and Leckie, J.O. (1982) Effect of complexation by Cl, SO_4 and S_2O_3 on adsorption of Cd on oxide surfaces. *Environmental Science and Technology*, **16** (3), 162-169.

Benjamin, M.M. (1983) Adsorption and surface precipitation of metals on amorphous iron oxyhydroxides. *Environmental Science and Technology*, **17** (11), 686-691.

Berner, R.A. and Rao, Ji-Long (1994) Phosphorus in sediments of Amazon river and estuary: Implications for the global flux of phosphorus to the sea. *Geochim. Cosmochim. Acta*, **58** (10), 2333-2339.

Braester, C. and Martinell, R. (1988) The Vyredox and Nitredox methods of in situ treatment of groundwater. *Water Science and Technology*, **20** (3), 149-163.

Broberg, O. (1984) Phosphorus removal in acidified and limed lake. *Water Research*, **18** (10), 1273-1278.

Cleasby, J.L. (1975) Iron and Manganese Removal - A case study. *Journal AWWA*, **67** (3), 147-149.

Dzombak, D.A. and Morel, F.M.M. (1990) *Surface Complexation Modelling - Hydrous Ferric Oxides.* John Wiley & Sons.

Faust, S.D. and Aly, O.M. (1998) *Chemistry of Water Treatment.* Second Edition, Ann Arbor Press Inc., USA.

Graveland, A. and Heertjes, P.M. (1975) Removal of manganese from groundwater by heterogeneous autocatalytic oxidation. *Transaction of Institution of Chemical Engineers*, **53**, 154-164.

Gu, B., Schmitt, J., Chen, Z., Liang, L. and McCarthy, J.F. (1994) Adsorption and desorption of natural organic matter on iron oxide: mechanisms and models. *Environmental Science and Technology*, **28** (1), 38-46.

Gu, B., Schmitt, J. and McCarthy, J.F. (1995) Adsorption and desorption of different organic matter fractions on iron oxide. *Geochim. Cosmochim. Acta*, **59**, 219-225

Hayes, K. F. and Leckie, J. O. (1987) Modelling ionic strength effects on cation adsorption at hydrous oxide/solution interfaces. *Journal of Colloid and Interface Science*, **115**, 564-572.

Hem, J.D. (1989) *Study and interpretation of the chemical characteristics of natural waters.* Third Edition. United States Geological Survey Water-Supply Paper 2254.

Hult, A. (1973a) Filtration of iron during and after oxidation. *Effluent and Water Treatment Journal*, April, 209–215.

Hult, A. (1973b) Removal of iron and manganese from a soft groundwater. *Effluent and Water Treatment Journal*, May, 293-296.

Knocke, W.R., Hamon, J.R. and Thompson, H. (1988) Soluble manganese removal on oxide coated filter media. *Journal AWWA*, **80** (12), 65-70.

Matthess, G. (1982) *The properties of groundwater.* Translated by Harvey J.C., John Wiley & Sons.

Meng, X. and Letterman, R. (1993) Effect of component oxide interaction on the adsorption properties of mixed oxides. *Environmental Science and Technology*, **27**, 970-975.

Morgan, J.J. (1964) *Chemistry of aqueous manganese II and IV.* Thesis, Harvard University, Cambridge, Massachusetts, USA.

O'Connor, J.T. (1971) Iron and Manganese. In: *Water Quality and Treatment - A Handbook of Public Water Supplies.* Chapter 11, pp 378-396. McGraw Hill, New York, USA.

Olson, L.L. and Twardowski Jr., C.J. (1975) $FeCO_3$ vs $Fe(OH)_3$ precipitation in water treatment plants. *Journal AWWA*, **67** (3), 150-153.

Robinson, L.R. Jr. and Breland, E.D. (1968) Removal of iron and manganese from low alkalinity waters. *Public Works,* **59** (2), 72-76.

Schenk, J.E. and Weber Jr., W.J. (1968) Chemical interaction of dissolved silica with iron(II) and iron(III). *Journal AWWA*, **60** (2), 199-212.

Schwertmann, U. and Fetcher, H. (1982) The point of zero charge of natural and synthetic ferrihydrites and its relation to adsorbed silica. *Clay Minerals*, **17**, 471-476.

Schwertmann, U. and Thalman, H. (1976) The influence of [Fe(II)], [Si] and pH on the formation of lepidocrocite and ferrihydrite during oxidation of aqueous $FeCl_2$ solutions. *Clay Minerals*, **11**, 189-200.

Snoeyink, V.L. and Jenkins, D. (1980) *Water Chemistry.* John Wiley & Sons, USA.

Standard Methods for the Examination of Water and Wastewater (1995) 19th edition, American Public Health Association/ American Water Works Association/ and Water Environment Federation, Washington D.C., USA.

Stumm, W. and Morgan, J.J. (1981) *Aquatic Chemistry - An introduction emphasizing on chemical equilibria in natural waters.* Second Edition, John Wiley & Sons Inc., USA.

Stumm, W. and Morgan, J.J. (1996) *Aquatic Chemistry - Chemical equilibria and rates in natural waters*: Third Edition, John Wiley & Sons Inc., USA.

Swallow, K.C., Hume, D. N. and Morel, F.M.M. (1980) Sorption of copper and lead by hydrous ferric oxide. *Environmental Science and Technology*, **14**, 1326-1331.

Takai, T. (1973) Studies on the mechanisms of catalytic deferrization (III) - Clarification on the mechanics of selfcatalytic deferrization. *Journal of Japan Water Works Association*, No 467, August, 16-23.

Tamura, H., Goto, K. and Nagayama, M. (1976) The effect of ferric hydroxide on the oxygenation of ferrous ions in neutral solutions. *Corrosion Science*, **16**, 197-207.

Tipping, E. (1981) The adsorption of aquatic humic substances by iron oxide. *Geochim. Cosmochim. Acta*, **45**, 191-199.

Tipping, E. and Cooke, D. (1982) The effects of adsorbed humic substances on the surface charge of goethite (∀-FeOOH) in fresh waters. *Geochim. Cosmochim. Acta*, **46**, 75-80.

Vaithiyanathan, P. and Corell, D.L. (1992) The Rhode river watershed: Phosphorus distribution and export in forest and agricultural soil. *Journal of Environmental Quality*, **21** (2), 280-288.

Weber, W.J. Jr. (1972) *Physicochemical processes for water quality control.* Wiley Interscience Publication; John Wiley & Sons Inc., USA.

Weber, W.J., Jr. (1985) Adsorption theory, concepts and models. In: *Adsorption Technology - A Step by Step Approach to Process Evaluation and Application.* Slejko, F.L. (Editor), Chapter 1, 1-36. Marcel Dekker Inc., New York, USA.

Yang, R.T. (1972) *Gas separation by adsorption processes.* Imperial College Press, London, UK.

Chapter 4

CHARACTERISATION OF COATED SAND FROM IRON REMOVAL PLANTS

Part of this Chapter is a paper by S.K. Sharma, B. Petrusevski and J.C. Schippers (2001) *Proceedings of 2nd IWA World Water Congress (15-19 October 2001), Berlin, Germany.*

ABSTRACT

Process efficiencies of iron removal plants increase with the development of iron oxide coatings on the filter media. These coatings have high adsorption capacities for iron(II), which facilitate the adsorptive iron removal in filters. However, iron oxide coatings developed under different conditions may have different physicochemical characteristics and, hence different adsorption capacities. The physical and surface chemical characteristics of coated sand from twelve different groundwater treatment plants in the Netherlands were analysed to assess their potential use in adsorptive iron removal. Specific surface areas, amounts of iron, manganese, calcium, and TOC in the coatings and iron(II) adsorption capacities of the coated media were measured. Additionally, grain size distribution and density of the media were determined and X-ray diffractograms, scanning electron micrographs, and EDAX spectra were prepared. Compared to new sand, coated sand had a very high porosity and a very large specific surface area. The higher the iron loading, the higher was the surface extractable iron content of the coated sand. The iron content of the coatings ranged from 27% to 45%. At pH 6.5, the iron(II) adsorption capacities of different coated sands were 10 to 55 times that of new sand. In general, iron(II) adsorption capacities of the coated sand from different plants increased with the increase in the time in use and the iron content of the coating. However, the average annual increase of iron content of the coatings and the iron adsorption capacity were different for the coated sands from different plants, likely due to the difference in water quality, process conditions applied, and time in use. The grain size of the filter sand increased and their density decreased with the development of the coating. The decrease in density was a function of the increase in the effective grain size. EDAX analysis showed that iron and oxygen were the main elements in the coating followed by manganese, calcium, silicon and carbon. All the coatings analysed were X-ray amorphous or poorly crystalline. It was observed that the iron oxide coating developed on the filter media was not uniform and different portions of the coated sand grains could have different elemental composition of the coating and surface characteristics. The measured high adsorption capacities of coated sand from wet filters and dry filters of full-scale groundwater treatment plants indicate that, in wet filters, adsorptive iron removal also plays a role. In dry filters, this mechanism should be dominant due to a very short pre-oxidation time. Process efficiencies of wet filters with iron oxide coated sand can be further improved by operating the filters predominantly in the adsorptive filtration mode.

4.1 INTRODUCTION

Iron oxide coatings develop on the surface of the filter media used in iron removal plants with time. This indicates that in groundwater treatment plants, adsorptive iron removal plays a role as well. Iron removal efficiencies are expected to increase with the development such coatings (Hauer 1950; Cox 1964; Stevenson 1997) since these coated media have high iron(II) adsorption capacity (see Chapter 2). This high adsorption capacity can be utilised to improve the efficiency of the existing iron removal filters with iron oxide coated sand by changing the mode of operation from conventional oxidation-floc formation to adsorption-oxidation. However, iron oxide coating developed on the filter media under different water quality (e.g. pH, iron concentration, HCO_3^-) and process conditions (e.g. time in use, pre-oxidation time, backwash conditions) may have different physical and chemical characteristics (Lo *et al.* 1997) and hence different iron(II) adsorption capacities. No information is, however, available on the elemental composition of the coating, specific surface area and the adsorption properties of the iron oxide coated sand from iron removal plants. Furthermore, the role of these coatings in iron removal is not yet fully understood. Hence, the objective of this study is to analyse the physical and surface chemical characteristics of the iron oxide coated sand developed under different conditions in twelve full-scale groundwater treatment plants in the Netherlands to assess their potential use in adsorptive iron removal.

4.2 THEORETICAL BACKGROUND

4.2.1 Iron oxide coated sand and iron removal

Iron oxide coating development on filter media in iron removal plants is a natural process in that the coatings develop in situ within weeks to months on new or virgin filter media during normal filter operation. The coatings thus developed consequently may reduce the ripening time after backwashing and improve the iron removal efficiency of a filter (Cox 1964; O'Connor 1971; Anderson *et al.* 1973; Stevenson 1997). Additionally, a broad range of commonly non-monitored (possibly unknown) improvements in the filtered water quality may result from the presence of such coatings as iron (hydr)oxide can adsorb various uncomplexed and complexed heavy metals, other inorganic anions and organic compounds (Bailey *et al.* 1992; Benjamin *et al.* 1996; Joshi and Chaudhuri 1996; Chang *et al.* 1997; Korshin *et al.* 1997). For efficient iron removal through the adsorption-oxidation mechanism, it is essential that coatings developed on the filter media have high iron(II) adsorption capacity. In addition, process conditions applied should be such that the pre-oxidation of iron before entering the filter is minimal e.g. as in dry filters.

4.2.2 Characteristics of the filter media coatings

The physicochemical properties of the filter media, especially the specific surface area and surface chemistry (type of functional groups present, nature of adsorption sites, point of zero charge, zeta potential and surface charge) have profound effects on both rate and capacity of adsorption. (Weber 1972, 1985). Furthermore, the physical and surface chemical characteristics of the filter media change with the development of the iron oxide coating. Specific surface area of the filter media increases with the development of coating (Chang *et al.* 1997; Lo *et al.* 1997). This is one of reasons why the adsorption capacity of iron oxide coated media is much higher than that of new (virgin) media.

The iron(II) adsorption capacity of the coated media depends, among others, on the concentration of iron (hydr)oxide at the surface of the filter media and the mineralogy (elemental composition, structure and properties) of the coating. The iron (hydr)oxides formed under different water quality conditions may have different chemical compositions and crystal structures with different specific surface areas, surface site densities and adsorption characteristics (Fasiska 1967; Crosby *et al.* 1983). Secondly, iron (hydr)oxides age over time to form stable end products. It is known that different iron oxides have different iron(II) adsorption capacities (see Chapter 2). Sung and Morgan (1980) reported that the initial products of iron(II) oxidation are lepidocrocite (γ-FeOOH) and amorphous FeOOH. These forms are unstable and age over time to form goethite (α-FeOOH).

Furthermore, water quality and the treatment schemes used in the plant may also have a significant effect on the nature of the coating developed on the filter media. Iron oxide can adsorb a host of cations, anions (Dzombak and Morel 1990; Kanungo 1994; Benjamin *et al.* 1996) and water-borne humic substances (Tipping 1981; Gu *et al.* 1994, 1995). The formation, morphology, transformation and adsorption capacity of iron (hydr)oxide is readily influenced or modified by the presence or adsorption of these "foreign" species (Cornell and Schwertmann 1996). Stenkamp and Benjamin (1994) found that the surface characteristics of underlying sand could also have some effect on those of coated sand, specifically if the coating is thin or porous. Therefore, the surface characteristics of coated filter media are physically and chemically heterogeneous and are expected to change with time.

4.2.3 Change in grain size and density of the media

The development of oxide coating increases the grain size (Galvin 1982; Reijnen 1984) and decreases the density (Prasad and Belsare 1984) of the filter media. From practice, it is known that in several groundwater treatment plants there are substantial "losses" of filter media during backwashing. The extent of these losses varies considerably among different plants. These losses can be attributed to the growth of the filter media, which reduces the depth of supernatant and the free board in the filter. Differences in filter media losses among different plants indicate differences in the growth rate of the coating. Additionally, the media growth and consequent change in density of the grain may affect the head loss developments and the efficiency of filter backwashing. Ultimately, the filter media need to be cleaned or replaced. An understanding of the factors affecting the growth of filter media is useful to assess the frequency of cleaning or replacement of the filter media and to improve filter design by incorporating the expected changes in freeboard, depth of supernatant and backwashing.

4.3 MATERIALS AND METHODS

In order to study the physical and surface chemical characteristics of iron oxide coated sand developed under different water quality and process conditions, iron oxide coated sand from twelve groundwater treatment plants in the Netherlands were collected and the filter media coatings were analysed. The major quality parameters of the raw water treated at these water treatment plants (WTP) are presented in Appendix 4.1. Of the twelve plants, WTPs Helden, Lichtvoorde and Sellingen apply dry filtration, whereas all others apply wet filtration.

Specific surface areas, surface extractable iron, manganese, calcium and TOC contents and iron(II) adsorption capacities of the coated media were measured. Additionally, grain size distribution and density were determined and X-ray diffractograms, scanning electron micrographs (SEMs), and Energy Dispersive Analysis of X-ray (EDAX) spectra were prepared.

Specific surface area and pore volume: Specific surface areas of the filter media were measured using the Brunauer, Emmett, and Teller (BET) nitrogen gas adsorption method (relative pressure range 0.05-0.25) with a Quantachrome Autosorb-6B gas adsorption analyser at 77 K at the Department of Chemical Technology of Delft University of Technology. Before surface area measurement, the samples were vacuum treated for 40 hours at 60°C. Additionally, pore volumes of some selected samples were also determined from N_2 adsorption isotherms.

Chemical extraction: Surface extractable iron, manganese, calcium and TOC contents of the media coatings were determined by chemical extraction. Replicate samples of about 2 to 4 grams of the coated sand were extracted in acidified hydroxylamine chloride ($NH_2OH.HCl$) at pH <2. Subsequently, the extracts were filtered and analysed by flame atomic absorption spectrometry (Perkin Elmer AAS 3110) and the amount of iron, manganese and calcium in the coating per gram of the filter media were quantified (Standard Methods 1995). The TOC content of the extract was measured using Model 700 TOC Analyser of O-I Corporation Texas, USA. The loss in weight of the media after extraction was determined to quantify the total amount of coating and the fraction of different elements in the coating.

Measurement of iron(II) adsorption capacity: Iron(II) adsorption capacities of the coated sand were measured by conducting a series of batch adsorption equilibrium experiments. The experimental set-up and the procedure were similar to that used in the earlier study (see Chapter 2). Experiments were performed at a pH of 6.5∀0.05 at room temperature (19 ± 1°C) using HCO_3^--CO_2 buffer. Different amounts of 400 mg/l deoxygenated iron(II) stock solution were introduced into the closed reactor containing a known weight of coated sand in deoxygenated water. On equilibrium, the iron concentration in the reactor was measured (Standard Methods 1995) and the iron(II) adsorbed onto the media was calculated from mass balance. The Freundlich isotherm model was fitted to the adsorption equilibrium data and isotherm constants K and n were obtained. For comparison of the adsorption capacities of coated sand of different sizes, the isotherm constant K was recalculated in terms of amount of iron(II) adsorbed per unit geometric surface area of the filter sand.

Grain size and density: The grain size distribution and the dry density of the coated media were measured before and after the extraction of the coating to quantify the change in grain size distribution and density with the development of the coating (Ives 1990; Degremont 1991).

Media surface analysis: The iron oxide coatings on the surface of the filter media were abraded mechanically. Powder X-ray diffractions (XRD) were conducted, at the Laboratory of Material Science, Delft University of Technology, using Siemens D500 goniometer (anode CuKα kV-mA 45-30) to analyse the mineralogy (chemical composition and crystal structure) of media coatings from different groundwater treatment plants. X-ray diffractogram is a plot of observed diffracted X-ray intensity against the Bragg angle θ (at which X-ray enters and leaves the crystal). Each crystalline phase or mineral can be identified from its characteristic X-ray diffraction pattern.

SEMs of the new (virgin) and coated sand were prepared, at the Laboratory of Material Science, Delft University of Technology, using JOEL Scanning Microscope (JSM-6400F) for the analysis of surface features and composition of the coatings. Filter media samples were air dried, glue mounted and gold coated by vacuum electric arc before scanning. Quantitative elemental analyses were performed with an EDAX system attached to the SEM to determine the elemental compositions of the surface mineral deposits on the media. The operating conditions were as follows: live time 100 seconds, dead time 35 seconds, operating voltage 10-25 kV, area analysed 25 x 25 micron.

4.4 RESULTS AND DISCUSSION

4.4.1 Specific surface area and pore volume

The BET specific surface areas of the coated filter sand from different plants are presented in Table 4.1. These specific surface areas are 3400-221000 times larger than their nominal geometric surface areas calculated from the sieve analysis data for the equivalent spherical grains (See Appendix 2.1). Measurements showed that pore volumes and specific surface areas of the filter media increase considerably with the development of iron oxide coating. The coated sand from WTP Gilze had an average pore volume of 1.16×10^{-2} cm^3/g and the BET surface area of 11.4 m^2/g, whereas the new sand of the same size had an average pore volume of 0.13×10^{-2} cm^3/g and the BET surface area of 1.0 m^2/g. Among the filter media tested, coated sand from WTP Noord Bargares had the highest specific surface area of 201 m^2/g with an average pore volume of 14×10^{-2} cm^3/g. These results reveal that compared to the new filter media, the iron oxide coated media have a much higher porosity and larger specific surface area and, hence are likely to have a higher iron(II) adsorption capacity.

Table 4.1 Characteristics of the coated sand tested

Coated sand from water treatment plant	Time in use years	Grain Size mm	Specific surface area m^2/g	Isotherm constants K mg/m^2	n -
WTP1 Roosendaal, WNWB	0.6	1.4 - 2.0	5.4	23.6	0.77
WTP2 Gilze, WNWB	3.0	0.7 - 1.25	11.4	61.1	0.54
WTP3 Helden, WML	4.0	1.4 - 2.0	34.0	36.8	0.67
WTP4 Ossendrecht, DELTA	7.1	1.4 - 2.0	46.5	69.3	0.38
WTP5 Oosterhout, WNWB	6.0	0.8 - 1.25	8.7	58.3	0.55
WTP6 Boxmeer, WOB	15.3	1.0 - 2.0	177.0	87.3	0.69
WTP7 Vlijmen, WOB	7.2	1.0 - 1.6	10.0	50.3	0.54
WTP8 Macharen, WOB	1.0	1.4 - 2.5	32.0	46.4	0.61
WTP9 N Bargeres, WMD	9.3	1.8 - 4.0	201.0	91.6	0.77
WTP10 Lichtvoorde, WG	1.5	0.8 - 2.0	6.5	49.2	0.95
WTP11 Brucht, WMO	7.5	1.12 - 2.24	33.0	80.2	0.73
WTP12 Sellingen, WGr	1.5	2.0 - 4.0	108.0	136	0.66

4.4.2 Chemical extraction

Surface extractable iron, manganese, calcium and TOC contents, and the loss of weight on extraction for different coated sand are summarised in the Table 4.2.

Table 4.2 Surface extractable contents of different coated sand

Coated sand from water treatment plant	Surface Extractable Content				Weight of coating mg/g sand	% of iron on coating
	Iron mg/g sand	Manganese mg/g sand	Calcium mg/g sand	TOC mg/g sand		
WTP1	15 ± 1	0.2 ± 0.0	0.3 ± 0.0	0.9 ± 0.1	62 ± 19	27 ± 8
WTP2	30 ± 1	4.6 ± 0.3	1.5 ± 0.2	5.7 ± 0.3	79 ± 6	39 ± 3
WTP3	90 ± 4	0.1 ± 0.0	0.1 ± 0.0	2.5 ± 0.2	219 ± 37	42 ± 6
WTP4	84 ± 3	8.0 ± 0.4	0.5 ± 0.0	3.5 ± 0.2	204 ± 5	42 ± 2
WTP5	45 ± 1	2.7 ± 0.1	1.8 ± 0.2	4.1 ± 0.3	124 ± 7	36 ± 2
WTP6	320 ± 4	49.7 ± 2.6	10.7 ± 0.2	31.7 ± 1.9	908 ± 6	35 ± 1
WTP7	161 ± 2	0.4 ± 0.1	4.9 ± 0.1	25.8 ± 1.2	436 ± 3	37 ± 1
WTP8	70 ± 3	0.9 ± 0.1	1.6 ± 0.3	2.9 ± 0.3	158 ± 17	45 ± 6
WTP9	330 ± 7	15 ± 1.6	5.7 ± 0.8	21.5 ± 2.8	821 ± 17	40 ± 1
WTP10	136 ± 5	1.4 ± 0.1	7.1 ± 0.5	26.0 ± 3.3	365 ± 9	37 ± 1
WTP11	257 ± 4	0.7 ± 0.1	7.0 ± 0.6	13.1 ± 1.0	728 ± 9	35 ± 1
WTP12	173 ± 4	0.2 ± 0.0	1.8 ± 0.5	7.4 ± 0.5	460 ± 13	38 ± 1

The surface extractable iron content (SEIC) can be used as a relative measure of the amount of surface iron oxide coating present on the media. SEIC of the media coatings were in the range of 27% to 45%. The fractions of manganese, calcium and TOC in the coating ranged from 0.04% to 5.8%, 0.05% to 1.95% and 0.11% to 7.22%, respectively. This indicated a wide variation in the chemical composition of the coatings. The loss of weight of different coated sand tested on extraction of the coating (or the total weight of coating) ranged from 6% to 91%. This shows that the extent of media growth and, consequently, the thickness of the iron oxide coating also varied widely.

Figure 4.1 shows the general trend that the SEIC of the coated sand was higher when the iron loading (time in use x raw water iron concentration) was higher. Higher SEIC in turn resulted in higher BET specific surface area of the coated sand. It was also found that the higher the manganese, calcium and TOC content of the raw water, the higher were their fractions in the coating.

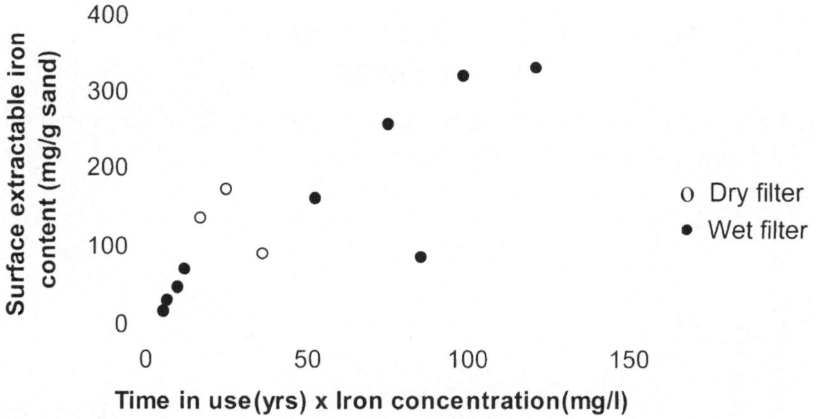

Fig. 4.1 Variation of SEIC with iron loading (time in use x raw water iron concentration)

4.4.3 Iron(II) adsorption capacity

The Freundlich isotherm constants K and n for iron(II) adsorption onto different coated sand at pH 6.5 are presented in Table 4.1. The iron(II) adsorption capacity of different iron oxide coated sand varied widely. The iron(II) adsorption isotherm constant K for new sand at a pH of 6.5 is 2.5 mg/m^2 (Chapter 2). Depending on the type of coating, the iron(II) adsorption capacity per unit geometric surface area of the different types of iron oxide coated sands were 10 to 55 times that of new sand. The measured high adsorption capacities of coated sand from wet filters and dry filters of full-scale groundwater treatment plants indicate that, in wet filters, adsorptive iron removal also plays a role. In dry filters, this mechanism should be dominant due to a very short pre-oxidation time. For all the coated sands tested adsorption isotherms were favourable ($n < 1$) or concave downwards.

Figure 4.2 presents the iron(II) adsorption capacity of coated sand from different plants as a function of time in use. In general, the adsorption capacity increased with the time in use and an increase in the iron content of the coating. However, the average annual increase of the iron content and the adsorption capacity was different for the coated sand from different plants. This can be attributed to the difference in water quality, process conditions and time in use.

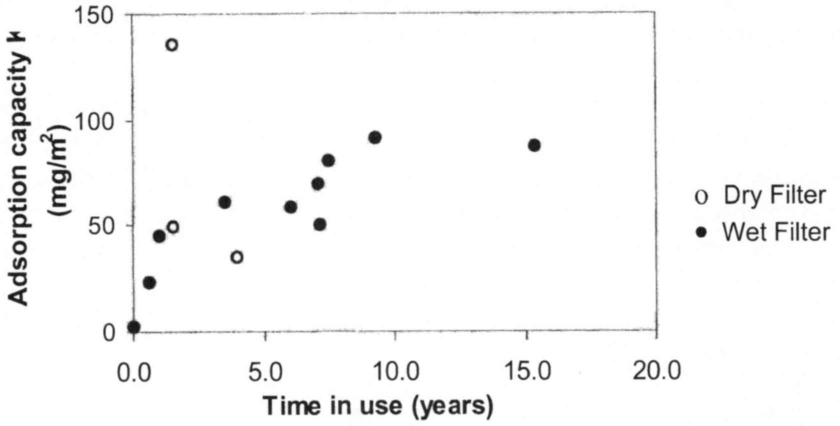

Fig. 4.2 Iron(II) adsorption capacity (isotherm constant K) of the coated sand from different plants as a function of time in use

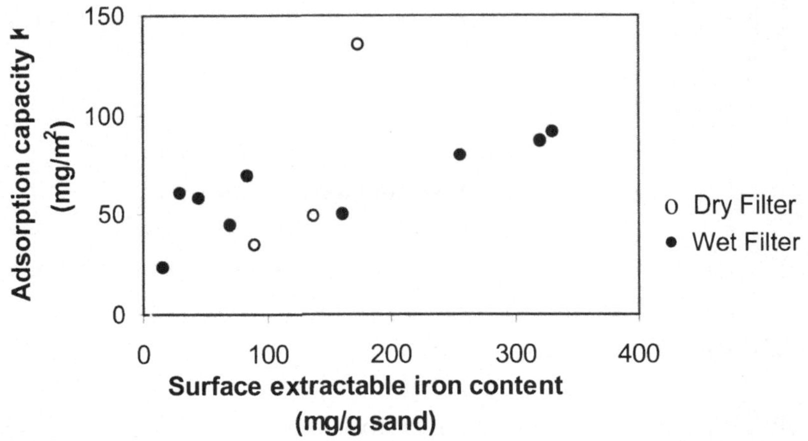

Fig. 4.3 Effect of SEIC on iron(II) adsorption capacity (isotherm constant K) of coated sand from different plants

There was no linear correlation between the iron(II) adsorption capacity and the SEIC (Fig. 4.3). This is probably because the density of iron(II) adsorption sites depends not only on the iron content and the specific surface area, but also on the chemical composition, crystal structure and surface properties of the coating. Secondly, there can be a significant difference between the total surface area of an adsorbent (as measured by the BET method) and the actual surface area available for a particular adsorbate. This is related to the sizes and distribution of the pores within the adsorbent particle. It is very likely that a portion of the total available surface area is inaccessible to the iron(II) ions (Weber 1985) due to blocking and/or filling of the pores with newly deposited iron hydroxides. No direct relationship was found between the raw water pH and the adsorption capacity of the coated sand developed at that pH.

4.4.4 Grain size and density

Table 4.3 presents effective grain sizes (d_{10}) and densities of the coated sand before and after the extraction of the coating. The grain sizes of filter sand from iron removal plants had increased with time in use. In some WTPs filter grains were growing very fast (e.g. WTPs Lichtvoorde, Sellingen), whereas in others the growth was relatively slow (WTPs Gilze, Helden, Oosterhout). The average annual growth rate of the effective grain size varied from 0.7% to 20.7%/year. Sieve analysis showed that regardless of the extent of increase in the grain size of the media from different WTPs with coating development, there was no significant change in the uniformity coefficient (d_{60}/d_{10}) of the filter grains. This indicates that there was a uniform growth of the filter grains in a given treatment plant suggesting that the filter sand was properly mixed during backwashing. This is likely the case because backwashing with air and/or air and water is commonly applied.

Table 4.3 Effective sizes and densities of coated and extracted sand

Coated sand	Effective size (d_{10})			Density		
from water treatment Plant	Coated sand mm	Extracted sand mm	Average growth rate %/year	Coated Sand Kg/m^3	Extracted sand kg/m^3	Type of rapid sand filter
WTP1	1.43	1.42	1.2	2590 ± 30	2560 ± 30	Wet
WTP2	0.80	0.78	0.7	2580 ± 20	2560 ± 20	Wet
WTP3	1.47	1.43	0.7	2610 ± 20	2560 ± 10	Dry
WTP5	0.75	0.72	0.7	2560 ± 10	2530 ± 10	Wet
WTP6	2.26	0.94	9.2	2030 ± 30	2530 ± 10	Wet
WTP7	1.33	1.08	3.2	2430 ± 50	2530 ± 15	Wet
WTP8	1.81	1.76	2.8	2580 ± 30	2530 ± 15	Wet
WTP9	2.23	1.20	9.2	2360 ± 30	2560 ± 40	Wet
WTP10	1.14	0.87	20.7	2410 ± 10	2570 ± 10	Dry
WTP11	1.51	0.91	8.8	2310 ± 30	2560 ± 40	Wet
WTP12	2.52	1.99	17.8	2340 ± 10	2550 ± 50	Dry

The coating development or the growth of the filter media was much faster in the case of dry filters (WTPs Sellingen, and Lichtvoorde). This is likely because in the case of dry filters pre-oxidation time is minimal and more iron(II) is available for adsorption. However, in the case of WTP Helden, the growth rate of the filter media was low even when they were from the dry filter probably due to low raw water pH. Additionally, it was observed that the coating development was also faster in the case of groundwater with high concentrations of iron and organic matter (WTPs Boxmeer, Noord Bargares and Brucht). Organic matter present in water can adsorb on the surface of the filter media. Secondly, organic matters can form complexes with iron and calcium. Therefore, it is likely that organic matter adsorbs on the surface of the media and creates more sites for the adsorption of iron and calcium, thus resulting in the rapid growth of the media. The effect of raw water quality on the rate of growth of filter media in iron removal plants needs further detailed investigation.

It was observed that when the iron oxide coating was relatively thin (coating constituting <20% of total volume or weight of the coated sand), there was no significant change in the density of the sand with the coating development. However, with an increase in coating thickness, the density of the grain decreased considerably, probably due to the development of a porous coating. Prasad and Belsare (1984) also reported the decrease in density of filter sand with the development of oxide coating. Figure 4.4 shows that the decrease in density of the coated sand was a function of the increase in the effective grain size (d_{10}). The increase in grain size and consequent decrease in density of the coated sand implies that the tendency of the "grown" grains to move towards the bottom of the filter bed during backwashing is reduced and filter beds would probably be fully mixed even after many years of operation.

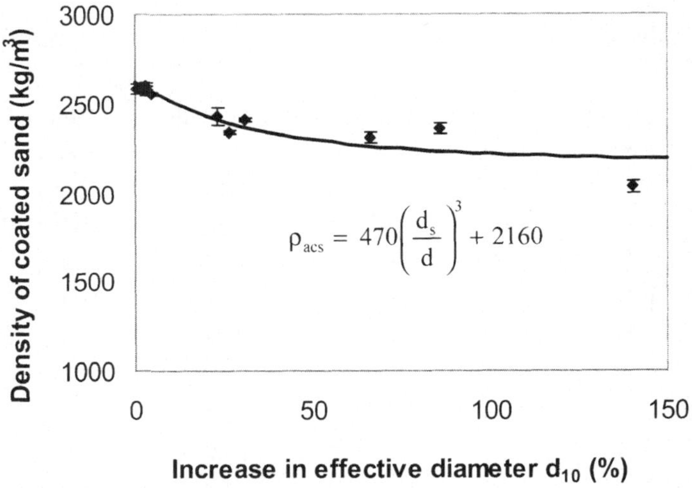

Fig. 4.4 Decrease in density of the coated sand from different plants with increase in effective grain size

Theoretically, the relation between densities of coated sand, new sand and iron oxide coating is given by

$$\frac{\pi d^3}{6}\rho_{acs} = \frac{\pi d_s^{\,3}}{6}\rho_s + \frac{\pi}{6}(d^3 - d_s^{\,3})\rho_c \qquad (4.1)$$

where ρ_{acs} = average density of the coated sand (kg/m^3)
ρ_s = average density of the new sand (kg/m^3)
ρ_c = average density of the iron oxide coating (kg/m^3)
d = average diameter of the coated sand (m)
d_s = average diameter of the core sand grain, without coating (m)

On simplification of equation (4.1) we have,

$$\rho_{acs} = (\rho_s - \rho_c)\left(\frac{d_s}{d}\right)^3 + \rho_c \tag{4.2}$$

This shows that the average density of coated sand is a function of the increase in grain size, density of sand and density of the iron oxide coating. As the filter media grows bigger with the development of the coating, the density of the coated sand approaches the density of the coating.

Since the efficiency of the adsorptive iron removal process primarily depends on the iron(II) adsorption capacity (which increases substantially with the coating development), the increase in grain size is not likely to affect the removal efficiency. However, with the increase in grain size and decrease in density of filter sand due to coating development, at the given backwash rate the filter bed expansion will be lower. This may not be critical in the case of adsorptive filters, in which backwashing is mainly intended for the regeneration of the adsorption sites (intermittent regeneration mode) and removal of minor flocs formed, if any (continuous regeneration mode). However, for filters operating under conventional oxidation-floc formation mode, in the long-term, an increase in grain size may impair the filter efficiency considerably. Secondly, since the bed depth also increases with the growth of the filter media, a part of the finer fraction of the filter media will probably be lost during backwashing.

The effect of a change in grain size and density with the coating development is likely to be more pronounced in the case of dual media filters with anthracite and sand. The average density of the iron oxide coatings developed on sand from different WTPs was calculated to be in the range of 2000-2800 kg/m^3. The dry density of anthracite is about 1400-1500 kg/m^3. Therefore, it is likely that when the anthracite grain is coated with iron oxide, its density will increase. At the given backwash rate, the expansion of the filter bed will consequently be reduced and the intermixing of anthracite and sand layers can be expected, due to an alteration of density and grain sizes.

4.4.5 Media surface analysis

X-ray diffraction: Powder X-ray diffraction (XRD) of the media coatings of sand from different plants showed that there were no distinct iron oxide crystals present on the coatings. The X-ray diffractograms of coatings did not match the standard X-ray diffraction pattern of any of the known crystalline minerals, indicating that the coatings were amorphous (X-ray amorphous) or very poorly crystalline. It is likely that the filter media coatings consist of a mixture of poorly ordered iron oxides together with some built-in impurities. X-ray diffractograms of all the coated sand were very similar and overlapping except that of the coated sand from WTP Helden, which had some small peaks indicating poor crystallinity. X-ray diffraction indicated that the iron oxide on the coating of sand from WTP Helden was predominantly maghemite (γ–Fe_2O_3). Typical diffractograms of coated sand from WTP Helden and WTP Boxmeer are shown in Fig. 4.5.

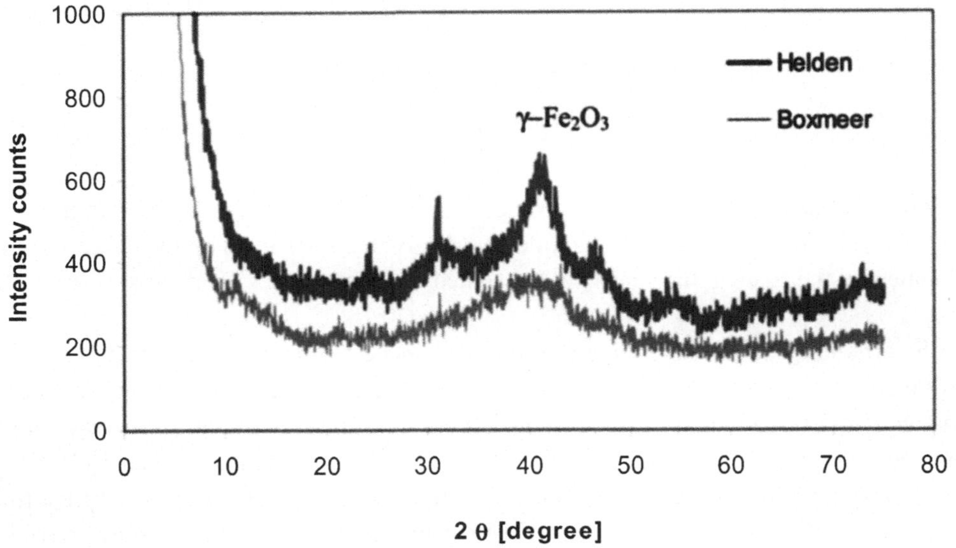

Fig. 4.5 X-ray diffractograms of coated sand from WTP Helden and WTP Boxmeer

Iron, oxygen, manganese, calcium, silicon and carbon were the main elements found in the coating. The presence of silica favours the formation of amorphous iron oxides (Robinson *et al.* 1981; Carlson and Schwertmann 1996; Mayer and Jarrell 1996). Huang and Wang (1997) reported that adsorption of silica, phosphate or organic matter on iron (hydr)oxide inhibits or retards its natural transformation to more crystalline or stable forms. Manganese oxides also modify the crystallisation process of iron oxides, resulting in the formation of X-ray amorphous and/or different crystalline iron oxides. This can be an explanation for the iron oxide coating on filter sand being very poorly crystalline or X-ray amorphous.

Scanning Electron Micrographs and EDAX analysis: SEMs of the filter media coatings were prepared to get an insight of the nature of the coating. SEMs of the surfaces of new sand and coated sand from WTP Gilze are presented in Figs. 4.6 and 4.7 respectively. Figure 4.7 shows some cracks on the surface of the coating, which may have already existed on the coating or might have developed during the preparation of the samples for SEM analysis. SEMs of the coated sand revealed nodules and globular forms on the coating and there was a high degree of surface roughness (Figs. 4.8 and 4.9). These micrographs showed that coated sand has a larger surface area (Figs. 4.8 and 4.9) compared to new sand (Fig. 4.6).

Fig. 4.6 SEM of new sand surface

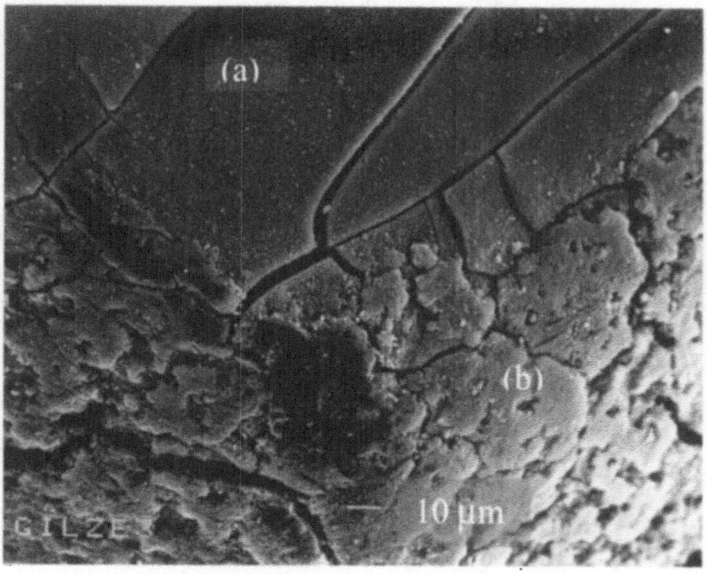

Fig. 4.7 SEM of the surface of the coated sand from WTP Gilze showing two different surface features (a) smooth region and (b) rough region

Fig. 4.8 SEM of smooth region of the surface of the coated sand from WTP Gilze at a higher magnification

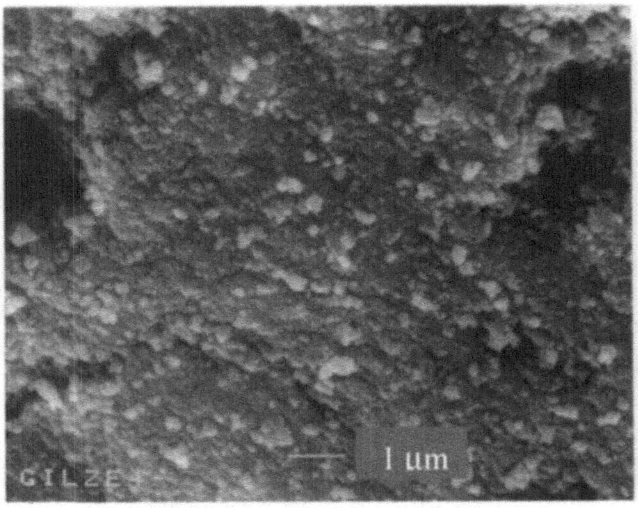

Fig. 4.9 SEM of rough region of the surface of the coated sand from WTP Gilze at a higher magnification

EDAX analysis revealed that iron and oxygen were the two major elements of all the coatings tested. Other elements present in the coating were manganese, calcium, silicon and carbon. This shows that, in addition to iron, different ions present in the raw groundwater like manganese, calcium, silicate, and organic matter are also adsorbed on the iron oxide coating. The adsorption of these elements influences the mineralogy and, hence, probably alters the adsorption capacity of the coating and influences the regeneration the iron(II) adsorption sites.

Two different types of surface features, namely "smooth region" (a) and "rough region" (b), were observed on the SEM of the coated sand from WTP Gilze (Fig. 4.7). EDAX analysis showed that there was a significant difference in the elemental composition of the coating in these two regions. The region with the smooth surface was mainly iron oxide with some silicon and calcium but no manganese (Figs. 4.8 and 4.10), whereas the region with the rough surface was composed of iron and manganese oxides together with silicon and calcium (Figs. 4.9 and 4.11). This shows that iron oxide coating developed on the filter media was not uniform throughout the filter grain and different regions of the coated sand could have different elemental composition and different surface characteristics and, hence, different adsorption properties. Consequently, it is most likely that process conditions can be adapted in such a way that adsorption capacity of the coating will become much larger. In Chapter 5, it is demonstrated that pH plays an important role. More detailed investigation might reveal major improvements in the adsorption capacity e.g. the role of manganese, acid treatment of the coating etc.

Gold peaks observed in the spectrum were from the gold coating applied to the filter media before SEM analysis.

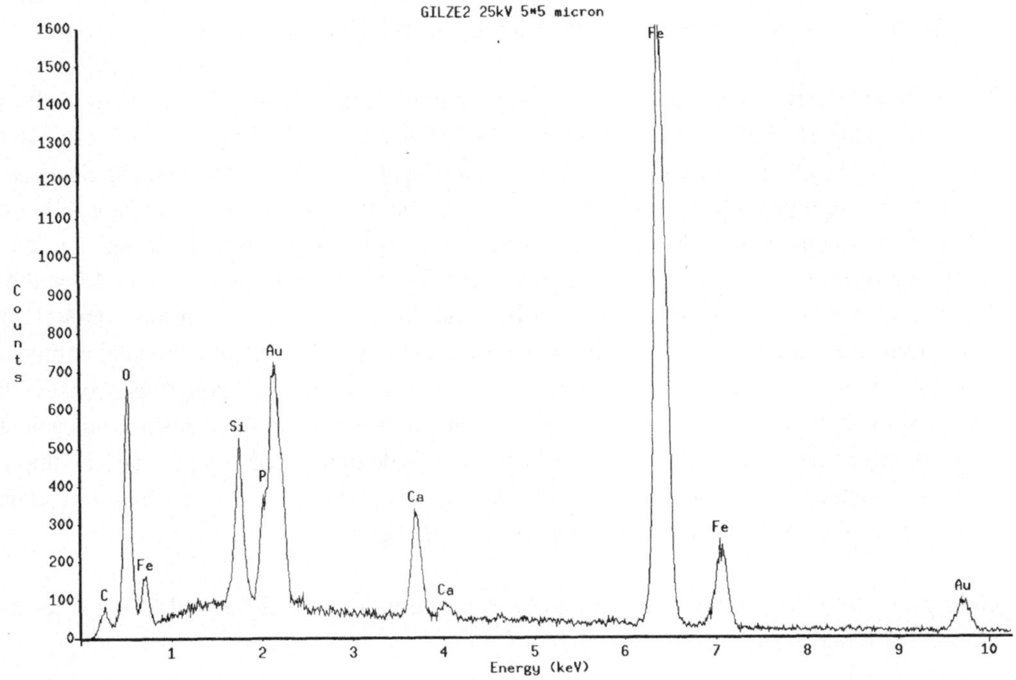

Fig. 4.10 EDAX spectrum of smooth region of the surface of the coated sand from WTP Gilze

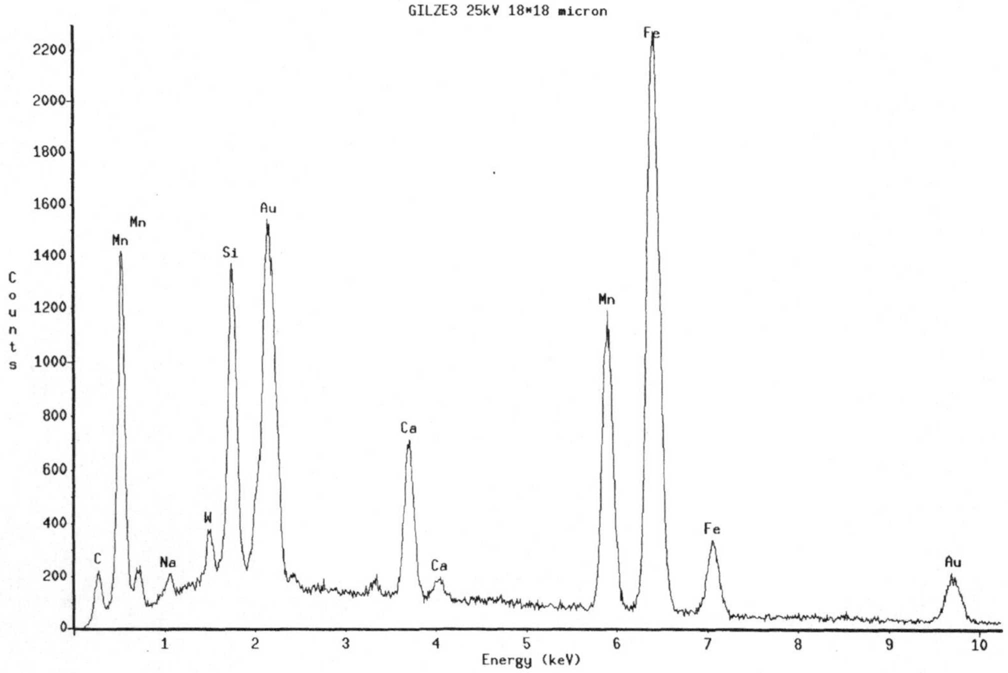

Fig. 4.11 EDAX spectrum of rough region of the surface of the coated sand from WTP Gilze

4.5 CONCLUSIONS

Based on the analysis of the physical and chemical characteristics of iron oxide coated sand from twelve iron removal plants in the Netherlands, the following conclusions can be drawn:

- Compared to new sand, iron oxide coated sand had a very high porosity (up to 110 times) and a very large specific surface area (5-200 times).
- The higher the iron loading (time in use x raw water iron concentration), the higher was the surface extractable iron content of the coatings. The iron content of the coatings ranged from 27% to 45%, indicating a substantial difference in chemical composition of the coatings.
- At pH 6.5, the adsorption capacities of different coated sands were 10 to 55 times that of new sand. In general, iron(II) adsorption capacity was higher for the coated sand with a high surface extractable iron content and large specific surface area. However, the average annual increase of the iron content and the adsorption capacity varied for the coated sand from different plants, probably due to the difference in water quality, process conditions, and time in use.
- The grain size of the filter sand increased and the density decreased with the development of iron oxide coating. The decrease in density of coated sand with the iron oxide coating development was a function of the increase in the effective grain size.
- Iron oxides on the coatings were X-ray amorphous or poorly crystalline. SEM/EDAX analysis showed that iron and oxygen were the main components of the coating and elements like manganese, calcium, silicon, and carbon were also present in the coatings. Analysis of coated sand from WTP Gilze showed that the coating developed on a filter grain was not uniform and different regions of the coating on a sand grain could have a different elemental composition and surface characteristics.
- The measured high adsorption capacities of coated sand from wet filters and dry filters of full-scale groundwater treatment plants indicate that, in wet filters, adsorptive iron removal also plays a role. In dry filters, this mechanism should be dominant due to a very short pre-oxidation time.
- As all the iron oxide coated sand tested demonstrated high iron(II) adsorption capacity, it is expected that process efficiencies of wet filters with coated sand can be further improved by operating the filters predominantly in the adsorptive filtration mode.

REFERENCES

Anderson, D.R., Row, D.D. and Sindelar, G.E. (1973) Iron and manganese studies of Nebraska water supplies. *Journal AWWA*, **65**, 637-641.

Bailey, R.P, Bennett, T. and Benjamin, M.M. (1992) Sorption onto and recovery of Cr(VI) using iron oxide coated sand. *Water Science and Technology*, **26** (5-6), 1239-1244.

Benjamin, M.M., Sletten, R.S., Bailey, R.P. and Bennett, T. (1996) Sorption and filtration of metals using iron oxide coated sand. *Water Research*, **30** (11), 2609-2620.

Carlson, L. and Schwertmann, U. (1987) Iron and manganese oxides in Finnish groundwater treatment plants. *Water Research*, **21** (2), 65-170.

Chang, Y., Li, C.W. and Benjamin, M.M. (1997) Iron oxide coated media for NOM sorption and particulate filtration. *Journal AWWA*, **89** (5), 100-113.

Cornell, R.M. and Schwertmann, U. (1996) *The Iron Oxides- structure, properties, reactions, occurrence and uses.* Published jointly by VCH- Germany and USA.

Cox, C.R. (1964) Iron and Manganese Removal. In: *Operation and Control of Water Treatment Processes.* Chapter 12, pp 209- 218, WHO, Geneva.

Crosby, S. A. *et al.* (1983). Surface areas and porosities of Fe(III)- and Fe(II)-derived oxyhydroxides. *Environmental Science and Technology*, **17** (12), 709-713.

Degremont (1991) *Water Treatment Handbook.* Vol. 1 and 2, Sixth Edition, Degremont, France.

Dzombak, D.A. and Morel, F.M.M. (1990) *Surface Complexation Modelling - Hydrous Ferric Oxides.* John Wiley & Sons, Inc., USA.

Fasiska, E.J. (1967) Structural aspects of the oxides and hydroxides of iron. *Corrosion Science*, **7**, 833-839.

Galvin, R.M. (1992) Ripening of silica sand used for filtration. *Water Research*, **26** (5), 683-688.

Gu, B., Schmitt, J., Chen, Z., Liang, L. and McCarthy, J.F. (1994) Adsorption and desorption of natural organic matter on iron oxide: mechanisms and models. *Environmental Science and Technology*, **28** (1), 38-46.

Gu, B., Schmitt, J. and McCarthy, J.F. (1995) Adsorption and desorption of different organic matter fractions on iron oxide. *Geochim. Cosmochim. Acta*, **59**, 219-225

Hauer, G.E. (1950) Iron and carbon dioxide removal. *Journal AWWA*, **42**, 555-561.

Huang, P.M. and Wang, M.K. (1997) Formation chemistry and selected surface properties of iron oxides. In: *Soils and Environment - Soil Processes from Mineral to Landscape scale.* Auerswald, K., Stanjek, H and Bigham, J.M. (Editors), Advances in Geoecology 30, 241-270. Catena Verlag Gmbh, Germany.

Ives, K.J. (1990) Testing of filter media. *Journal of Water SRT-Aqua*, **39** (3), 144-151.

Joshi, A. and Chaudhuri, M. (1996) Removal of arsenic from groundwater by iron oxide coated sand. *Journal of Environmental Engineering*, **122** (8), 769-771.

Kanungo, S.B. (1994) Adsorption of cation on hydrous iron oxides: I. Interfacial behaviour of amorphous FeOOH and β-FeOOH (Akaganeite) in different electrolyte solution. *Journal of Colloid and Interface Science*, **162**, 86-92.

Korshin, G.V., Benjamin, M.M. and Sletten, R.S. (1997) Adsorption of natural organic matter on iron oxide: Effects of NOM composition and formation of organo-halide compounds during chlorination. *Water Research*, **31**(7), 1643-1650.

Lo, S.L., Jeng, H.T. and Lai, C.H. (1997) Characteristics and adsorption properties of iron-coated sand. *Water Science and Technology*, **35** (7), 63-70.

Mayer, T.D. and Jarrell, W.M. (1996) Formation and stability of iron(II) oxidation products under the natural concentrations of dissolved silica. *Water Research*, **30** (5), 1208-1214.

O'Connor, J.T. (1971) Iron and Manganese. In: *Water Quality and Treatment - A Handbook of Public Water Supplies.* Chapter 11, pp 378-396. McGraw Hill Book Company, New York, USA.

Prasad, D.Y. and Belsare, D. K. (1984) The physico-chemical characteristics of used filter sand and its role in the seven water systems in a tropical region. *Acta Hydrochim. et Hydrobiol.*, **12** (1), 73-80.

Reijnen, G.K. (1984) Optimaliseren van de ontijzering. *H_2O*, **17** (21), 478-481.

Robinson, R.B., Demirel, T. and Baumann, E.R. (1981) Identity and character of iron precipitates. *Journal of Environmental Engineering*, **107** (EE6), 1211-1227.

Standard Methods for the Examination of Water and Wastewater (1995) 19th edition, American Public Health Association / American Water Works Association /Water Environment Federation, Washington, DC.

Stenkamp, V.S. and Benjamin, M.M. (1994) Effect of iron oxide coatings on sand filtration. *Journal AWWA*, **86** (8), 37-50.

Stevenson, D.G. (1997) *Water Treatment Unit Processes.* World Scientific Publishing Company.

Sung, W. and Morgan, J.J. (1980) Kinetics and products of ferrous iron oxygenation in aquatic systems. *Environmental Science and Technology*, **14** (5), 561-568.

Tipping, E. (1981) The adsorption of aquatic humic substances by iron oxide. *Geochim. Cosmochim. Acta*, **45**, 191-199.

Weber, W.J., Jr. (1972) *Physicochemical processes for water quality control.* Wiley Interscience Publication; John Wiley & Sons Inc., USA.

Weber, W.J., Jr. (1985) Adsorption theory, concepts and models. In: *Adsorption Technology - A Step by Step Approach to Process Evaluation and Application.* Slejko, F.L. (Editor), Chapter 1, 1-36. Marcel Dekker Inc., New York, USA.

APPENDIX 4.1 Major raw water quality parameters of groundwater treatment plants

Parameter	Units	Roosendaal WNWB	Gilze WNWB	Helden WML	Ossendrecht DELTA	Oosterhout WNWB	Boxmeer WOB	Vlijmen WOB	Macharen WOB	N Bargares WMD	Lichtvoorde WG	Brucht WMO	Sellingen WGr
pH	-	7.2	7.65	5.75	6.9	7.6	7.05	6.65	7.2	6.91	7.07	6.9	6.52
Fe	mg/l	7.2 -10	1.72	9	12	1.2 - 2.0	6.4	7.3	12	13.3	11.31	9.7	16.7
Mn	mg/l	0.2	0.07	0.35	0.16	0.05	1.1	0.34	1.5	0.49	0.53	0.39	0.3
Ca^{2+}	mg/l	63	52	30	44	74	113	44	98	50	158	69	39.2
Mg^{2+}	mg/l	4.5	4.3	13	4.2	7.9	12	3.5	6.8	6.5	10.2	5.4	4.4
Cl^{-}	mg/l	17	15	30	23	14.1	41	21	55	36	45	37	33
SO_4^{2-}	mg/l	10	10	120	50	5	120	29	44	57	81	25	0.8
HCO_3^{-}	mg/l	228	155	16	120	270	260	120	290	128	430	211	175
NH_4^{+}	mg/l	0.45	0.12	0.2	0.39	0.68	0.45	0.7	0.8	0.2	0.83	1.6	0.4
CH_4	mg/l	-	-	-	0.15	-	-	-	-	-	-	-	13
$KMnO_4$ cons	mg/l	11	4	1.5		7.1	16.5	13	4.5			16	
TOC	mg/l	3.7	1.5		1.9	2.11		3.9		2.0 (DOC)	6.1	3	5.25

Chapter 5

DEVELOPMENT OF IRON OXIDE COATING ON FILTER MEDIA

Part of this Chapter was published by S.K. Sharma, G. IJpelaar and J.C. Schippers (2001) *Transactions of The Filtration Society*, **1** (3), 68-72.

ABSTRACT

The development of an iron oxide coating on the filter media is an important factor in effective iron removal from groundwater. The rate of development of the coating and its characteristics may be influenced by the raw water quality, process conditions and characteristics of the filter media. Laboratory-scale short column experiments were conducted to study the effect of pH, influent iron concentration and type of filter media on the development of iron oxide coatings on the filter media. The increase in the surface extractable iron content (SEIC) and iron(II) adsorption capacity of filter media with the coating development was measured. Coating developed faster on filter sand at a pH of 6.5 and 7.0 than at a pH of 6.0. Coated sand developed at a pH of 6.5 and 7.0 had higher iron(II) adsorption capacity compared to coated sand developed at a pH of 6.0. Coating development was faster at the influent iron concentrations of 4.0 mg/l and 6.0 mg/l compared to 1.0 mg/l, and the coated sand developed at an iron concentration of 6 mg/l had the highest iron(II) adsorption capacity. Initially the iron(II) adsorption capacity increased rapidly with increasing SEIC. Subsequently, the iron(II) adsorption capacity increased at a much slower rate despite a steady increase in the SEIC, most likely because the sand surface was fully covered with iron oxide and the pores in the previously deposited layers were filled with new iron oxides. Among the three virgin (new) media tested, basalt had the highest iron(II) adsorption capacity, and the coating development was initially fastest on basalt followed by olivine and sand. Preconditioning of new media at higher feed water pH and/or higher iron concentration can be an attractive approach to reduce the initial filter ripening time and to rapidly develop media with a high iron(II) adsorption capacity.

5.1 INTRODUCTION

The conventional approach of iron removal is to oxidise the soluble iron(II) present in anoxic groundwater with oxygen in the air or some chemical oxidant and to remove the flocs formed in filters (O'Connor 1971; Salvato 1992). In several systems sedimentation is incorporated just before rapid sand filtration. Besides systems with wet filters, dry filtration is applied in groundwater treatment, particularly when ammonium is present as well.

Iron oxide coating develops on the filter media used in iron removal plants with time in use. Iron oxide coated sand has a high adsorption capacity for iron(II), which can be utilised to improve iron removal in filters (see Chapter 2). Filters with iron oxide coated media are expected to perform better in terms of filtrate quality and filter run time, if the governing iron removal mechanism is changed from conventional oxidation-floc formation (floc filtration) to adsorption-oxidation (adsorptive filtration).

The development of an iron oxide coating on filter media is an important factor for the effective iron removal from groundwater employing the adsorption-oxidation mechanism of iron removal.

The raw water quality, process conditions and characteristics of the filter media may influence the rate of development of the coating and its characteristics (surface area, surface charge, pore sizes and distribution, adsorption capacity) (Merkle *et al.* 1996). Iron oxide coatings developed under different conditions can have different physical and chemical characteristics (surface area, chemical composition, crystal structure and densities of adsorption sites) and hence different iron(II) adsorption capacities. Secondly, after the replacement of filter media in conventional groundwater treatment plants, the filters are initially "run to waste" until the media is fully "ripened". This initial ripening period ranges from a few days to one week for iron removal and a few weeks to three months for manganese removal (JMM 1985; Twort *et al.* 2000). During the ripening period, the iron and manganese oxide coating develops on new media, which aids in improving filtrate quality. Considerable savings are possible if the filter media coating can be developed faster. This will reduce the ripening period and increase the efficiency of the process. Therefore, for effective iron removal, it is important that the iron oxide coating develops rapidly on the filter media and exhibits a high adsorption capacity for iron(II). However, no information is available on the factors affecting the development of iron oxide coatings on the filter media and the effectiveness of coatings developed under different conditions for adsorptive iron removal.

Hence, the objective of this study is to analyse the effect of pH, influent iron concentration, and the type of filter media on the development of iron oxide coating on filter media in order to establish a procedure for the rapid development of coating with a high iron(II) adsorption capacity.

5.2 THEORETICAL BACKGROUND

There is growing evidence that iron hydroxides or iron-bearing materials have the capability of removing many metals over a wide pH range and with high efficiency (Benjamin 1983; Stenkamp and Benjamin 1994; Benjamin *et al.* 1996). Iron oxide coated sand has been successfully employed in fixed beds to treat metal-bearing water and wastewater (Edwards and Benjamin 1989; Bailey *et al.* 1992; Lai *et al.* 1994; Joshi and Chaudhuri 1996; Benjamin *et al.* 1996; Petrusevski *et al.* 2000). In most of the previous studies, iron oxide coated sand was prepared artificially by coating the filter sand with iron oxide chemically in the laboratory. This study focuses on the analysis of the process of development of iron oxide coating on the filter media during the normal filter operation in iron removal plants.

Processes of coating development

Theoretically, the iron oxide coating can develop on the filter media of iron removal plants by two main mechanisms:

(a) *Adsorption and Oxidation:* This mechanism involves adsorption of iron(II) onto the surface of the filter media and subsequent oxidation of adsorbed iron(II) in the presence of oxygen or other oxidant to form a coating. This is generally the case when the iron is entering the filter as iron(II) and removed by adsorption-oxidation.

(b) *Attachment and Consolidation:* Diz and Novak (1998) reported that iron oxide coating growth on the filter sand appeared to result from the attachment and consolidation of small iron particles (<1.0 μm) formed in the bulk solution. After oxidation and hydrolysis of iron(II), very fine particles of iron(III) hydroxides formed may be deposited on the surface of filter media. Some of these particles are not removed by backwashing and in the long term they are consolidated to form a coating on the surface of the filter media. This type of coating generally develops when iron reaches the surface of the filter media in iron(III) form.

Under conditions that favour adsorptive iron removal, iron(II) entering the filter is removed through adsorption onto the surface of the filter media. Hence, it is likely that the adsorption-oxidation mechanism will result in much faster coating development on filter media than the particle attachment-consolidation mechanism. However, under the conditions commonly applied in practice, both mechanisms are expected to take place simultaneously. Therefore, both mechanisms may contribute to the development of the iron oxide coating on the filter media.

Factors affecting iron oxide coating development

The development of iron oxide coating on the filter media in iron removal plants may be influenced by:

i) the groundwater quality (pH and concentrations of Fe^{2+}, Mn^{2+}, Ca^{2+}, TOC etc.);
ii) the process conditions applied (filtration rate, depth of supernatant, depth of the filter media, backwash conditions, aeration efficiency); and
iii) type and characteristics of the filter media (grain size, specific surface area).

It is well known that the pH affects the adsorption of ions present in water as well as the surface charge of hydrated surface of the filter media (Davis and Leckie 1978; Stenkamp and Benjamin 1994; Stumm and Morgan 1996). In general, high pH favours the adsorption of cations as it is coupled with the release of H^+ ion (Gadde and Laitinen 1974; Dzombak and Morel 1990). The iron(II) adsorption capacity of the filter media also increases with the increase of pH (see Chapter 2). Ibrahim (1997) reported that coating development (ripening) is faster at higher pH. The iron(II) adsorption capacity is the function of the equilibrium iron(II) concentration. Therefore, it is likely that coating development is faster in water with a high iron(II) concentration. The presence of other inorganic ions such as Ca^{2+}, Mn^{2+}, SO_4^{2-}, PO_4^{3-} and organic matter could also influence the iron oxide coating formation and its characteristics (see Chapter 3).

In general, new (virgin) media have a relatively lower adsorption capacity for iron(II), hence, it is likely that the coating development is slower at the beginning. The subsequent adsorption takes place on previously adsorbed and oxidised iron, which has a much higher adsorption capacity; thus, it is expected that the development of coating accelerates thereafter.

The depth of supernatant governs the time available for oxidation of iron(II) which, in turn, influences the amount of iron entering the filter bed in iron(II) form and hence its adsorption. For the given groundwater, it is very likely that iron oxide coating development is faster in wet filters with a low depth of supernatant and/or high filtration rate. It has also been observed in practice that filter media in dry filters develop iron oxide coating very quickly compared to filter media in conventional wet rapid filters. This is because in dry filters the pre-oxidation time is negligible. The transport of the iron(II) ions from the bulk solution to the surface of the filter media (coating) is mainly governed by the filtration rate. A high filtration rate, therefore, enhances the external mass transfer.

Filter backwashing conditions and specifically air scouring and high backwash rate could develop shear action that can disintegrate and wear off the iron oxide coating developed on the media. At the same time, backwashing (with air and water) provides ample opportunity for the filter media to intermix, thus promoting the uniform coating development along the entire depth of the media. Uniformity of the filter media favours the mixing of the filter media and the uniform coating development. In an intermittent regeneration mode of adsorptive iron removal, oxygen in backwash water is responsible for the oxidation of the adsorbed iron(II) and the regeneration of adsorption sites that is required for the continuity of the process. Therefore, optimisation of backwashing conditions could be equally important for the proper development of the iron oxide coating on the filter media.

The rate of iron oxide coating development and its characteristics are likely to be different for different filter media as they have different surface charge, specific surface area and different types and densities of adsorption sites. Stenkamp and Benjamin (1994) found that the surface characteristics of the underlying sand could also have some effect on those of coated sand, specifically if the coating is thin or porous. In general, it was observed that filter media with iron-bearing minerals like basalt and olivine have a high iron(II) adsorption capacity (see Chapter 2) and, consequently, may develop an iron oxide coating faster.

The adsorption-oxidation mechanism of iron removal relies on the iron(II) adsorption capacity of the filter media, which increases considerably after the development of iron oxide coating. Therefore, it is important that coating development on the filter media is faster. In some groundwaters the preconditioning of the new media may be necessary to speed up the ripening process. Such preconditioning may help to improve iron removal efficiency or to achieve higher efficiency even at the earlier stage of the use of new media. This could be an attractive approach for filter media replacement in many iron removal plants where iron oxide coated media have been in use for many years.

Simplified model of iron oxide coating development on filter media

Surface extractable iron content (SEIC) and iron(II) adsorption capacity of the filter media change with the development of the iron oxide coating on the surface of the filter media. The SEIC is the relative measure of the development of iron oxide coating on the filter media. The net increase in the SEIC of the filter media in one filter run (Δ SEIC) is the difference between the iron deposited on the media in a given filter run and the iron removed from the media during backwashing at the end of that filter run.

A. Total amount of iron deposited on the filter media in one filter run I_{dep} (g) is given by

$$I_{dep} = k_1 C V \tag{5.1}$$

where k_1 = deposition constant which depends on the overall iron removal efficiency (fraction of iron entering the filter bed in iron(II) form and the fraction of iron removed in the filter) (-)

C = influent iron concentration (g iron/m^3)

V = volume of water filtered during a run (m^3) = $v A \Delta t$

v = filtration rate (m/h)

A = area of the filter bed (m^2)

Δt = filter run time between two backwashings (h)

Therefore,

$$I_{dep} = k_1 v A C \Delta t \tag{5.2}$$

B. Total amount of iron removed from the filter media in one backwashing I_{rem} (g)

i) removal of iron deposited during a particular filter run is assumed to be proportional to the amount of iron deposited in that filter run

$= k_2 I_{dep} = k_2 k_1 v A C \Delta t$

where k_2 is a constant which depends on i) the mode of iron removal (adsorptive/floc formation or the fraction of iron entering the filter bed in iron(II) form) and ii) the scouring effect of backwashing conditions.

ii) removal of the iron from the coating also depends on the SEIC of the coating. Assuming that the removal of iron from the coating during one backwashing is equivalent to a continuous removal of iron from the coating during the filter run at a uniform rate R (g iron/g sand/h)

$R = k_3 SEIC$

where SEIC = surface extractable iron content at the beginning of a filter run (g iron/g sand), and k_3 = removal constant which depends on backwashing conditions.

Iron removed from the coating = $k_3 SEIC \Delta t M$

where M = total mass of the filter media in the filter (g sand)

Therefore,

$$I_{rem} = k_1 k_2 v A C \Delta t + k_3 SEIC \Delta t M \tag{5.3}$$

From mass balance of iron,

$$\Delta SEIC * M = I_{dep} - I_{rem} \tag{5.4}$$

From equations (5.2), (5.3) and (5.4) we have

$$\frac{\Delta SEIC}{\Delta t} = k_4 - k_3\,SEIC \tag{5.5}$$

where $k_4 = (k_1 - k_1 k_2)\,v\,A\,C / M$ = constant for given raw water and process conditions.

As ΔSEIC and Δt are very small compared to the SEIC of a coated sand and the total time required to develop an adequate coating, the rate of increase of SEIC with filter run time (t) can be approximated by

$$\frac{dSEIC}{dt} = k_4 - k_3\,SEIC \tag{5.6}$$

On integration and rearranging the terms we have,

$$SEIC = \frac{k_4}{k_3}(1 - e^{-k_3 t}) \tag{5.7}$$

Therefore, the SEIC increases asymptotically with time.

At the beginning of the filter run the SEIC of the coating is very small, therefore the removal of the coating during backwashing is expected to be independent of the SEIC ($k_3 = 0$). For this case, from equation (5.6) we have

$$SEIC = k_4 t \tag{5.8}$$

Therefore, it is anticipated that initially the SEIC will increase linearly with time, and when a sufficient coating develops and the filter media grows bigger an asymptotic increase of the SEIC with time will result.

The iron(II) adsorption capacity of filter media under given conditions (pH, grain size) is a function of the SEIC, the specific surface area, the thickness of coating, and the pore size distribution in the coating. It should be noted that iron(II) adsorption capacity is not directly proportional to the SEIC or the thickness of the coating because, with the development of new layers of the coating, the pore spaces of the internal layers are filled or partially blocked and therefore may not be accessible for iron(II) ions.

5.3 MATERIALS AND METHODS

5.3.1 Filter media

Characteristics of the filter media used in the experiments are summarised in Table 5.1. Specific surface areas of the filter media were measured using the BET nitrogen gas adsorption method (relative pressure range 0.05-0.25) with a Quantachrome Autosorb-6B gas adsorption analyser at 77 K at the Department of Chemical Technology of Delft University of Technology.

Table 5.1 Characteristics of the filter media used

Media	Grain size (mm)	d_{10} (mm)	UC (d_{60}/d_{10})	Geometric surface area (m^2/g)	BET specific surface area (m^2/g)
Sand	0.7 - 1.25	0.81	1.17	2.80 x 10^{-3}	0.9 ± 0.1
Olivine	0.8 - 1.4	0.86	1.36	2.40 x 10^{-3}	1.2 ± 0.1
Basalt	0.8 - 1.4	0.97	1.27	3.00 x 10^{-3}	1.4 ± 0.1

5.3.2 Filtration experimental set-up

The pilot filtration experimental set-up, as shown in Fig. 5.1, consisted of three 500 mm high plexiglass filter columns of 150 mm diameter. The initial depth of the filter media in each column was 150 mm in all experiments conducted in order to facilitate the uniform development of iron oxide coating. A model water was prepared by mixing $FeSO_4.7H_2O$ (Baker Analysed) with tap water (pH = 8.1, O_2 = 8-10 mg/l, HCO_3^- = 129 mg/l, Ca^{2+} = 53 mg/l, Mg^{2+} = 7.9 mg/l, Fe = 0.03 mg/l, Temp = 16°C) and adjusted to the desired pH with HCl (Merck). Iron(II) stock solution and HCl were dosed to the influent water and mixed before entering the filters. The filtration rate was maintained at 5 $\pm$ 0.5 m/h. Process conditions (depth of supernatant and contact time after introduction of iron(II)) were selected in such a way that in all filtration experiments more than 75% of the iron entering the filter bed was iron(II). After a run time of 48-60 hours, the filters were backwashed with water (50 m/h) for 10 minutes.

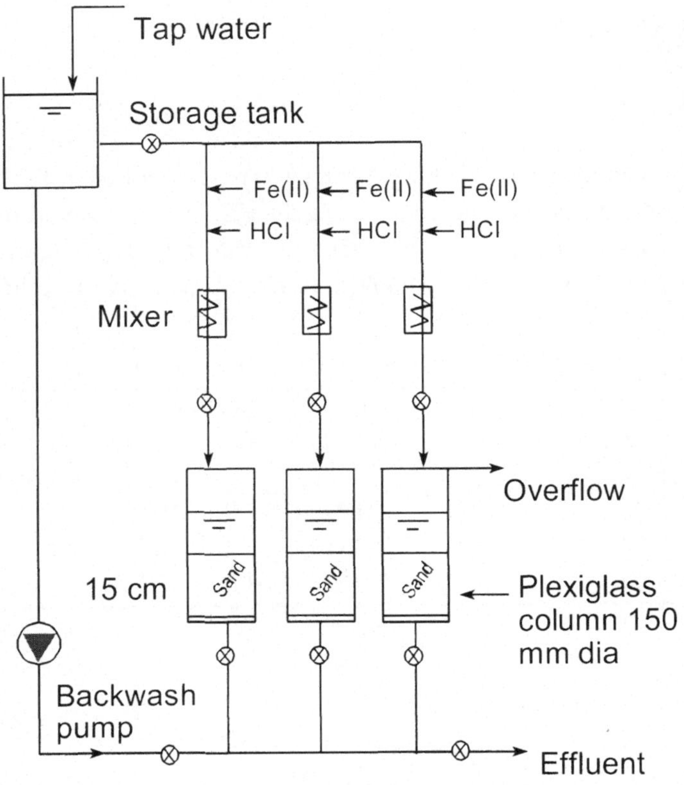

Fig. 5.1a Schematic of laboratory-scale column experimental set-up

Fig. 5.1b Photographs of laboratory-scale column experimental set-up

5.3.3 Experimental procedure

Iron oxide coatings were developed on the filter media by conducting filter runs with laboratory-scale columns at different pH (6.0, 6.5 and 7.0 ± 0.1), with different feed water iron concentrations (1.0, 4.0, and 6.0 ± 0.2 mg/l) and using three different media (silica sand, olivine and basalt). Filter media samples were taken periodically after backwashing the filters with tap water. For these samples, the surface extractable iron content (SEIC) and iron(II) adsorption capacity of the coated media were measured. The filtrate iron concentration was also monitored to assess the increase in removal efficiency with the coating development.

Surface extraction of the coated media
Replicate samples of 2-3 g of coated media were extracted with 20 ml of concentrated HCl. The iron concentrations of the extracts were measured using flame atomic absorption spectrometry (Perkin Elmer AAS 3110) (Standard Methods 1995).

Measurement of iron(II) adsorption capacity
A series of batch experiments were conducted with media samples taken periodically from the columns to determine their iron(II) adsorption capacity. The experimental set-up and procedure for the measurement of iron(II) adsorption capacity was similar to that used in earlier studies (see Chapter 2). Adsorption experiments were performed at a pH of 6.5 ± 0.05 and at room temperature (19 ± 1°C). 15 ml of 400 mg/l deoxygenated iron(II) stock solution was introduced

into the closed reactor containing about 50 g of filter media sample in 1.5 litres of deoxygenated water (initial iron(II) concentration ~ 4 mg/l). The equilibrium iron concentration in the reactor was measured (Standard Methods 1995) and the amount of iron(II) adsorbed onto the filter media was calculated from iron mass balance. The iron(II) adsorption capacity (Freundlich isotherm constant K) for each sample was then calculated from the Freundlich isotherm equation (eq. 2.5) using the n values of 0.73, 0.52 and 0.38 respectively for virgin sand, olivine and basalt and the n value of 0.6 for the coated media (see Chapter 2).

5.4 RESULTS AND DISCUSSION

5.4.1 Effect of pH

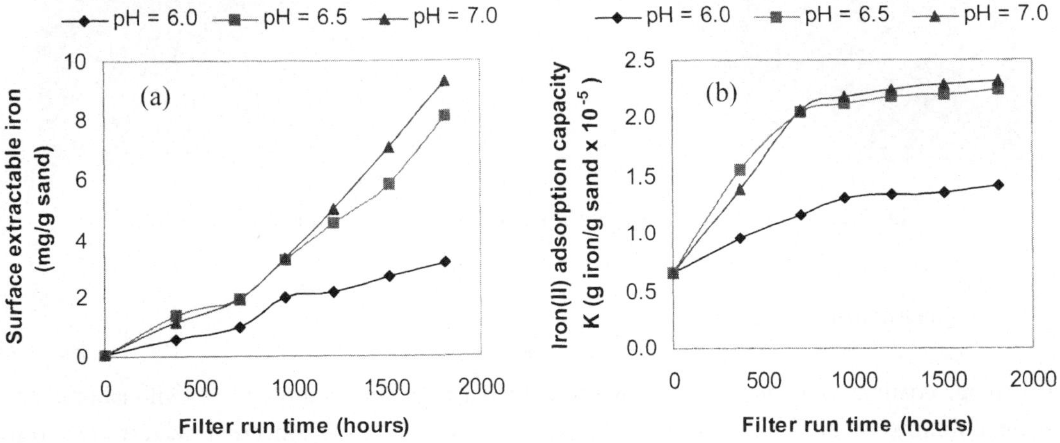

Fig. 5.2 Increase in (a) SEIC and (b) iron(II) adsorption capacity with time for coated sand developed at different pH (feed water iron concentration = 2.0 ± 0.2 mg/l, filtration rate = 5 ± 0.5 m/h)

Figure 5.2a shows that the SEIC increased with the filter run time and was higher for the sand coated at higher pH. This was attributed to the higher iron(II) adsorption capacity of sand at high pH. It was also found that the iron(II) adsorption capacity increased with the development of the coating (Fig. 5.2b). This is due to higher iron(II) adsorption capacity of iron oxide coated sand compared to new sand (see Chapter 2). As the iron oxide coating develops on the filter media, the specific surface area and the density of iron(II) adsorption sites increase and hence the iron(II) adsorption capacity increases. After a filter run time of 1820 hours, the iron(II) adsorption capacities of coated sand developed at influent pH of 6.0, 6.5 and 7.0 were respectively 210%, 340% and 350% of that of new (virgin) sand. Coated sand developed at pH of 6.5 and 7.0 exhibited considerably higher iron(II) adsorption capacity compared to coated sand developed at pH of 6.0 (Fig. 5.2b). This may be due to the difference in chemical composition, crystal structure and physicochemical properties of iron (hydr)oxides formed at different pH

(Cornell and Schwertmann 1996). Lo *et al.* (1997) also found that sand artificially coated with iron oxide at higher pH had a higher adsorption capacity for cadmium(II) and nickel(II).

Increase in removal efficiency with the coating development

The effect of coating development on the iron removal was assessed by monitoring the iron concentration in the filtrate during the coating development. Figure 5.3 shows the decrease in residual iron concentration, in samples taken one hour after backwashing, with the filter age at different pH values. The results obtained confirm that the iron removal efficiency increased with the coating development.

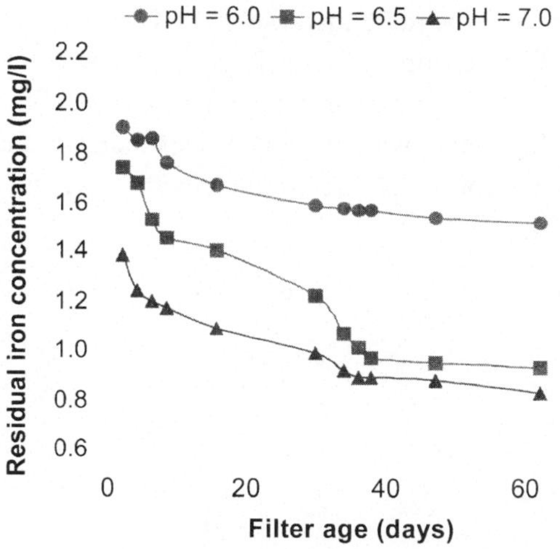

Fig. 5.3 Residual iron concentration of short filter columns with filter age at different pH (feed water iron concentration = 2 .0 ± 0.2 mg/l, filtration rate = 5.0 ± 0.5 m/h)

As the iron oxide coating develops on the filter media, more iron is expected to be removed by the adsorption-oxidation mechanism. Consequently, at pH 7.0 the decrease in residual iron concentration was most pronounced because of the higher iron(II) adsorption capacity of sand at higher pH (see Chapter 2). However, it is important to note that the pilot plant was not operated under anoxic conditions. Therefore, there was some oxidation of the iron(II) before it reached the top of the filter bed while some iron(II) oxidation also occurred in the filter bed. At a pH of 7.0, about 75-80% of the iron in the feed water was iron(II), while at a pH of 6.0 iron(II) was about 85-90%. Thus, during the filter runs, the removal of iron was taking place by both mechanisms, adsorption-oxidation as well as oxidation-floc formation. However, in all the three filters operated at different pH, adsorption-oxidation was the dominant removal mechanism as more than 75% of the iron was entering the filter bed in iron(II) form.

5.4.2 Effect of iron concentration

The change in SEIC and iron(II) adsorption capacity of coated sand developed at three different feed water iron concentrations and pH 7.1 as a function of filter run time are presented in Fig. 5.4a and 5.4b, respectively. Figure 5.4a shows that the SEIC increased with time in use and was higher for coated sand developed at higher influent iron concentrations. The iron(II) adsorption capacity increased with the development of the coating and was highest for coated sand developed at the highest iron concentration tested (6.0 mg/l) (Fig. 5.4b). After a filter run time of 1400 hours, the iron(II) adsorption capacities of coated sand developed at influent iron concentrations of 1 mg/l, 4 mg/l and 6 mg/l were respectively 380%, 670% and 800% of that of new (virgin) sand. After an initial rapid increase, iron(II) adsorption capacity increased slowly with the development of the coating. Figure 5.4b also shows that an increase in adsorption capacity of the coated sand was not proportional to an increase of influent iron concentration of the feed water in which the coating was developed. This indicates that at a given pH, there is an ultimate adsorption capacity for iron(II) determined by the type and density of adsorption sites.

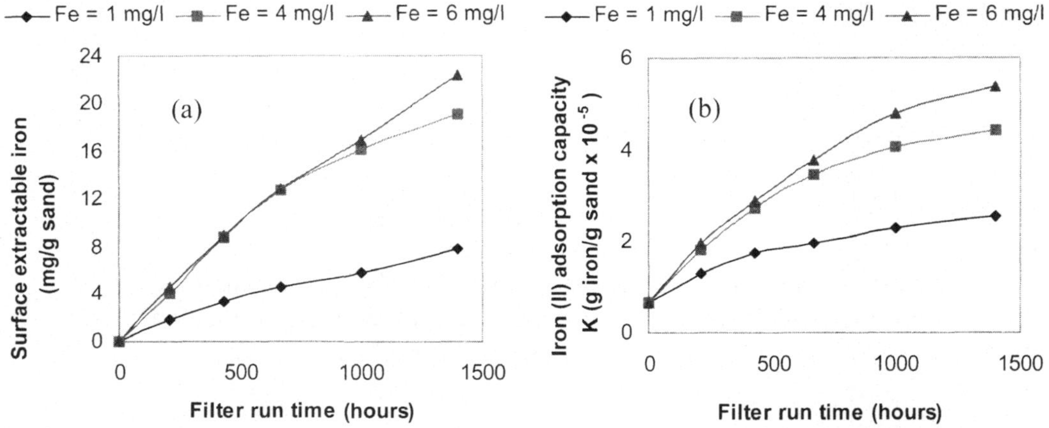

Fig. 5.4 Increase in (a) SEIC and (b) iron(II) adsorption capacity with time for coated sand developed at different iron concentrations (pH = 7.1 ± 0.1 filtration rate = 5 ± 0.5 m/h)

Figures 5.2 and 5.4 suggest that, irrespective of the feed water pH and iron concentration, initially both the SEIC and adsorption capacity increased rapidly with the development of the coating. Afterwards, the adsorption capacity continued to increase at a much slower rate despite a steady increase in the SEIC, most likely because the sand surface was fully covered with iron oxide and the pores in the previously deposited layers were filled or blocked.

In general, the more rapid the coating development (e.g. at higher pH or at higher iron concentration), the higher the iron(II) adsorption capacity. This is probably due to the development of an open iron (hydr)oxide structure that possesses a high specific surface area with more sites for iron(II) adsorption (Fasiska 1967; Crosby *et al.* 1983).

In water treatment practice the bed depth of filter media is between 1 to 2 m, which is much higher than that used in these experiments. As the iron concentration in water passing through the filter decreases along the depth of the filter, it is expected that coating development would be faster at the top layers of the filter compared to the bottom layers. Hence, it is likely that filter media at the upper layers of the filter have a higher iron(II) adsorption capacity compared to the media at the lower layers. In order to facilitate the uniform iron oxide coating development along the entire depth of the filter, it is therefore important that the media are properly mixed during backwashing.

Relation between SEIC and iron(II) adsorption capacity at given pH

Figure 5.5 shows that irrespective of the initial iron concentration, iron(II) adsorption increased with an increase in the SEIC. It also shows that the rate of increase of iron(II) adsorption capacity of the coated sand decreased with the coating development.

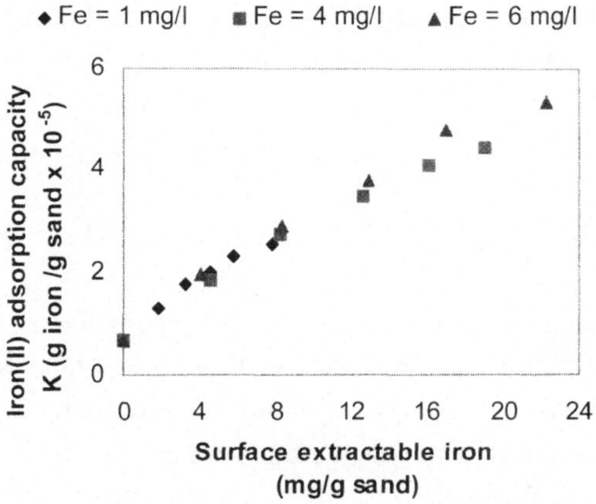

Fig. 5.5 Relationship between SEIC and iron(II) adsorbed for the coated sand developed at different feed water iron concentrations (pH = 7.1 ± 0.1, filtration rate = 5 ± 0.5 m/h)

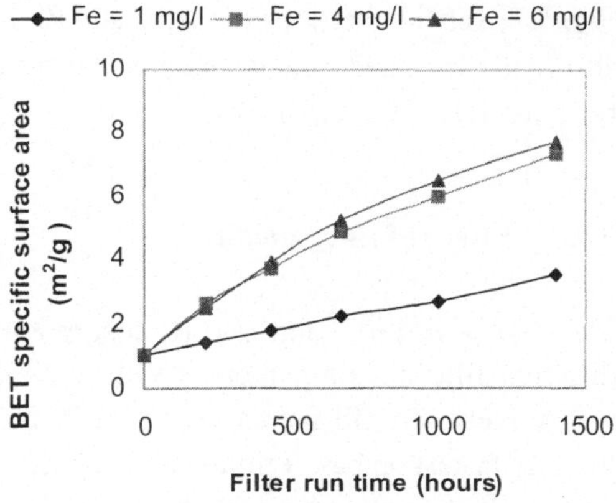

Fig. 5.6 Increase in the BET specific surface area with time for coated sand developed at different feed water iron concentrations (pH = 7.1 ± 0.1, filtration rate = 5 ± 0.5 m/h)

Relation between SEIC, BET specific surface area and iron(II) adsorption capacity

Figure 5.6 shows the increase in the BET specific surface area with filter run time for iron oxide coated sand developed at different feed water iron concentrations. It is clear from Fig. 5.6 that the BET surface area increased with the coating development and was higher for the coated sand developed at higher feed water iron concentrations.

Figure 5.7a shows that, at a given pH, there is a linear relationship between SEIC and BET surface area. Iron(II) adsorption capacity also increased with the increase in BET surface area (Fig. 5.7b).

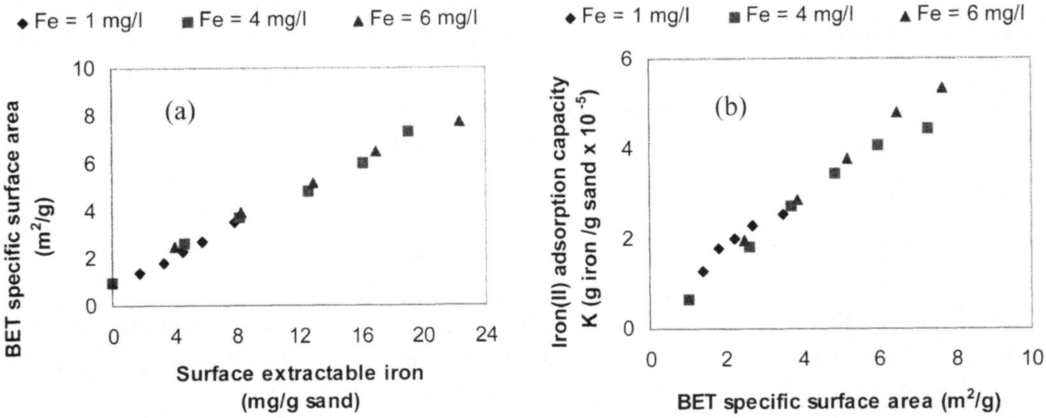

Fig. 5.7 Relation between (a) SEIC and BET surface area and (b) BET surface area and iron(II) adsorbed for coated sand developed at different feed water iron concentrations (pH = 7.1 ± 0.1 filtration rate = 5 ± 0.5 m/h)

5.4.3 Effect of filter media

The change in SEIC and iron(II) adsorption capacity with the coating development for three different filter media namely: basalt, olivine, and sand are presented in Figs. 5.8a and 5.8b respectively. For all media tested the SEIC increased with time. Coating development was initially fastest on basalt followed by olivine and sand. Afterwards the rate of increase of SEIC was nearly the same for all the three media. This is probably because after the complete coverage of the virgin media by iron oxide coating and filling of the internal pores, all the coated media would have similar surface characteristics. Virgin (new) basalt demonstrated very high iron(II) adsorption capacity that increased minimally with the coating development. Virgin sand and olivine had considerably lower iron(II) adsorption capacity, however, with the development of the coating, iron(II) adsorption capacity of both media increased rapidly.

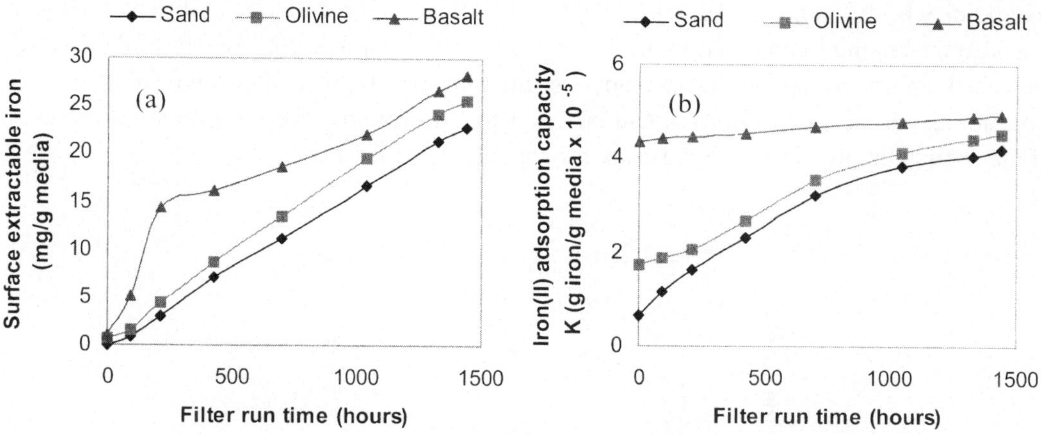

Fig. 5.8 Increase in (a) SEIC and (b) iron(II) adsorption capacity with run time for different filter media (pH = 7.1 ± 0.1, feed water iron concentration = 4.0 ± 0.2 mg/l, filtration rate = 5 ± 0.5 m/h)

After a filter run time of 1440 hours, the iron(II) adsorption capacities of virgin sand and olivine increased by 630% and 260% respectively, whereas the adsorption capacity of basalt increased only by 10%. It is expected that, under the same operating conditions, after a longer period of operation all the media tested will ultimately attain a very similar capacity for iron(II) adsorption.

Basalt is an igneous rock, dark in colour, weathering to brown and consists of iron and magnesium rich minerals like olivine $(Mg,Fe)_2SiO_4$ and augite $(Mg,Fe)O(Al,Fe)_2O_3SiO_2$. Iron oxides and iron-bearing minerals are known to adsorb iron(II) (Takai 1973; Tamura *et al.* 1976; Zhang 1997). The high iron(II) adsorption capacity of basalt is therefore likely to be due to the presence of iron-bearing minerals. It is likely that with the development of iron oxide coating there is no substantial change in the surface characteristics and adsorption properties of basalt, hence no significant increase in iron(II) adsorption capacity. Olivine is a common rock forming mineral containing iron and magnesium silicates. Consequently, the iron(II) adsorption capacity of virgin olivine is also much higher than that of virgin silica sand.

Experimental results show that virgin filter media with a high iron(II) adsorption capacity develops the iron oxide coating faster and hence is likely to have the shortest ripening time. Filters with such media do not need ripening time and can be used immediately without waiting for the coating development.

Figure 5.9 shows the increase in iron content (%) of the coating with run time for different filter media tested. After a run time of 1440 hours, the average iron contents on the coating of sand, olivine, and basalt were 50%, 48% and 33% respectively. The considerable difference in the iron content on the coating of basalt with that of sand and olivine indicates the difference in chemical composition and mineralogy of iron oxide coatings developed on different filter media. This is because, depending on the relative fractions of Fe, O and H present, the iron (hydr)oxide on the coating may be different (e.g. FeOOH, $Fe(OH)_3$, Fe_2O_3, Fe_3O_4 or a mixture of different oxides). It is also interesting to note that in the case of sand and basalt, the iron content of the coating increased rapidly during the first 200 hours of the filter run. Further filter run resulted in very slow additional increase of iron content in the coating. In contrast, the increase in iron content of the coating on olivine was very steady throughout the experiment.

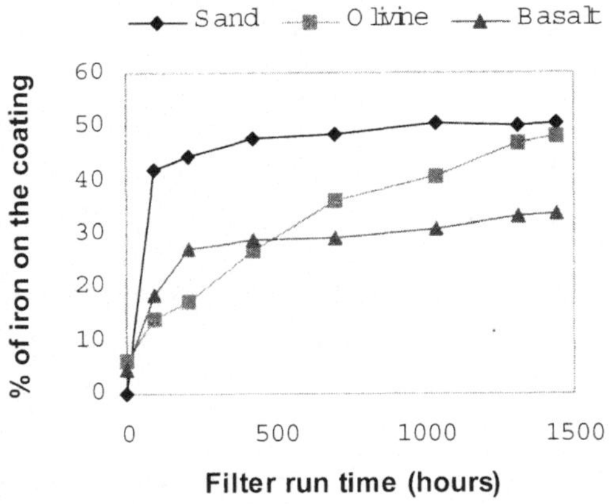

Fig. 5.9 Increase in % iron on the coating with filter run time for different filter media (pH = 7.1 ± 0.1, feed water iron concentration = 4.0 ± 0.2 mg/l, filtration rate = 5 ± 0.5 m/h)

The sieve analysis of the filter media samples before and after the development of the coating (under all three experimental conditions tested) showed that there was no significant increase in the grain size (≤10 percent along any axis). However, it is expected that in the long term the filter media will continue to grow in size due to iron(II) adsorption and subsequent oxidation on the surface of the media.

5.4.4 Implications for practice

Preconditioning of the new filter media

The experimental results imply that in water treatment practice, iron oxide coating development on new filter media could be facilitated by preconditioning of the filters which may include initial operation of the filters at higher pH and/or at high iron concentration together with regular backwashing to ensure proper mixing. Such preconditioning of new filter media can be an attractive approach to developing an iron oxide coating with an adequate iron(II) adsorption capacity more quickly. This is likely to reduce the initial ripening time of filters and improve the removal efficiency of conventional iron removal plants that are known to perform poorly in the initial phase after the replacement of filter media.

Batch adsorption experiments indicated that the SO_4^{2-} ion might enhance the development of iron oxide coating on new filter sand (see Chapter 3). However, this should be verified with short column tests.

Filter media preconditioning that involves continuous re-circulation of feed water with a high concentration of iron(II) sulphate under anoxic conditions with an intermittent backwashing with oxygen-rich water could be an alternate approach to speed up the development of iron oxide coating on filter media. Similar techniques have been used to coat filter media with iron and manganese oxides (JMM 1985; Knocke *et al.* 1988)

Additionally, establishing a preconditioning procedure for the rapid development of an iron oxide coated sand with a high adsorption capacity could be very useful in other water treatment areas, since recent studies have shown that iron oxide coated sand can successfully remove contaminants like arsenic, cadmium, chromium, copper, lead, nickel, selenium and zinc from water (Stahl and James 1991; Bailey *et al.* 1992; Lai *et al.* 1994; Joshi and Chaudhuri 1996; Benjamin *et al.* 1996; Viraraghavan and Li 1999; Petrusevski *et al.* 2000).

Filter media for adsorptive iron removal filters

Virgin filter media with a high iron(II) adsorption capacity develop the iron oxide coating faster and have a shorter ripening time. Consequently, filters with such media can be used immediately, as the time needed for the coating development is much reduced. This suggests that the use of filter media with a high iron(II) adsorption capacity, like basalt, in iron removal filters can be very attractive. However, other characteristics of such media should be tested before employing them in iron removal filters. There does not seem to be any hindrance to apply basalt as filter media as it is in use for a long term in Water Supply Gelderland in The Netherlands.

Stability of the media coating

The stable performance of coated filter media over a long period requires a sufficient amount of iron oxide to remain attached to the surface and to retain its adsorptive properties. In the pH ranges normally encountered in groundwater treatment practice (5-9), the dissolution of the coating is likely to be negligible. However, other ions and compounds present in the water may create a reductive condition consequently promoting the dissolution of the coating. Incorporation of other ions in the crystal lattice of iron oxides may also alter the iron(II) adsorption capacity of coated sand and the rate of regeneration of adsorption sites. Secondly, the disintegration and wearing off of the coating due to shear action during backwashing should be minimal. Hence, the backwashing procedure of the filter using coated media should allow removal of loose flocs accumulated in the filter bed without eroding the surface coating significantly.

The iron (hydr)oxides change their form and crystallinity with time (e.g. transformation of amorphous iron oxides and lepidocrocite to goethite). The ageing of the filter media coating may change the structure and properties of iron oxide and consequently influence the iron(II) adsorption capacity. It was found that the iron(II) adsorption capacity of lepidocrocite decreases with ageing (see Chapter 2). Hence, it is important that the iron oxide coating on the filter media is continuously regenerated so as to maintain the iron(II) adsorption capacity and to avoid a substantially negative effect of the ageing of the iron oxide coatings. During the treatment process in the longer run, however, the inner layer of the coating will gradually change its form and lose a part of the iron(II) adsorption capacity. This will, however, very likely be compensated by the newer layers of iron hydroxides formed on the surface of the filter media due to iron(II) adsorption and its subsequent oxidation. In order to verify whether the ageing of previously deposited iron oxides indeed plays a substantial role, further study should be conducted measuring iron(II) adsorption capacity as a function of time.

5.5 CONCLUSIONS

The process of development of iron oxide coating on the filter media was analysed by conducting laboratory-scale filter runs under different process conditions. Based on the experimental results the following conclusions can be drawn:

- The development of iron oxide coating on filter sand was faster at a pH of 6.5 and 7.0 than at a pH of 6.0. The coated sand developed at a pH of 6.5 and 7.0 had a considerably higher iron(II) adsorption capacity compared to coated sand developed at a pH of 6.0.
- Coating development was faster at feed water iron concentrations of 4.0 mg/l and 6.0 mg/l compared to 1.0 mg/l. The coated sand developed at the highest iron concentration tested (6 mg/l) had the highest iron(II) adsorption capacity. Initially, both the SEIC and the iron(II) adsorption capacity increased rapidly with the development of the coating. Further on, the adsorption capacity increased at a much slower rate despite a steady increase in the SEIC with time, most likely because the sand surface was fully covered with iron oxide and newly developed iron oxides covered and/or filled the pores in the previously deposited layers.
- Among the three media tested, virgin basalt had the highest iron(II) adsorption capacity, and the coating development was fastest on basalt, followed by olivine and sand. The use of filter media with a high natural iron(II) adsorption capacity, like basalt, in iron removal plants can be an attractive option to reduce the time required to develop a coating with adequate adsorption capacity and to improve the process efficiency
- Preconditioning of new filter media at high feed water pH and/or high iron concentration is expected to be an attractive approach to rapidly develop an effective iron oxide coating and to reduce the initial filter ripening time. An alternative approach would be to circulate water with a high iron(II) concentration under anoxic conditions with regular backwashing with oxygen-rich water.
- Further studies should be conducted to analyse the effect of backwashing conditions on iron oxide coating development and to examine the effect of ageing of the coating on the iron(II) adsorption capacity.

REFERENCES

Bailey, R.P, Bennett, T. and Benjamin, M.M. (1992) Sorption onto and recovery of Cr(VI) using iron oxide coated sand. *Water Science and Technology*, **26** (5-6), 1239-1244.

Benjamin, M.M. (1983) Adsorption and surface precipitation of metals on amorphous iron oxyhydroxides. *Environmental Science and Technology*, **17** (11), 686-691.

Benjamin, M.M., Sletten, R.S., Bailey, R.P. and Bennett, T. (1996) Sorption and filtration of metals using iron oxide coated sand. *Water Research*, **30** (11), 2609-2620.

Cornell, R.M. and Schwertmann, U. (1996) *The Iron Oxides- structure, properties, reactions, occurrence and uses.* VCH Publications.

Crosby, S. A. *et al.* (1983) Surface areas and porosities of Fe(III)- and Fe(II)-derived oxyhydroxides. *Environmental Science and Technology*, **17** (12), 709-713.

Davis, J.A. and Leckie, J.O. (1978) Surface ionization and complexation at the oxide/water interface: Surface properties of amorphous iron oxyhydroxide and adsorption of metal ions. *Journal of Colloid and Interface Science*, **67** (1), 90-107.

Diz, H.R. and Novak, J.T. (1998) Fluidized bed for removing iron and acidity from acid mine drainage. *Journal of Environmental Engineering*, **124** (8), 701-708.

Dzombak, D.A. and Morel, F.M.M. (1990) *Surface Complexation Modelling - Hydrous Ferric Oxides.* John Wiley & Sons, Inc.

Fasiska, E.J. (1967) Structural aspects of the oxides and oxyhydrates of iron. *Corrosion Science*, **7**, 833-839.

Edwards, M. and Benjamin, M.M. (1989) Adsorptive filtration using coated sand a new approach for the treatment of metal bearing wastes. *Journal of WPCF*, **61** (9), 1523-1533.

Gadde, R.R. and Laitinen, H.A. (1974) Studies of heavy metal adsorption by hydrous iron and manganese oxides. *Analytical Chemistry*, **46** (11), 2022-2026.

Ibrahim, A.K.M. (1997) *The development of iron oxide coating on filter media and its role in iron removal.* M.Sc. Thesis SEE 019, IHE - Delft, The Netherlands.

J.M.Montgomery Consulting Engineers (1985) *Water Treatment, Principles and Design.* John Wiley & Sons Inc., USA.

Joshi, A. and Chaudhuri, M. (1996) Removal of arsenic from groundwater by iron oxide coated sand. *Journal of Environmental Engineering*, **122** (8), 769-771.

Knocke, W.R., Hamon, J.R. and Thompson, H. (1988) Soluble manganese removal on oxide coated filter media. *Journal AWWA*, **80** (12), 65-70.

Lai, C.H., Lo, S.H. and Lin, C.F. (1994) Evaluating an iron oxide coated sand for removal of copper from water. *Water Science and Technology*, **30** (9), 175-182.

Lo, S.L., Jeng, H.T. and Lai, C.H. (1997) Characteristics and adsorption properties of iron-coated sand. *Water Science and Technology*, **35** (7), 63-70.

Merkle, P.B., Knocke, W., Junta-Ross, J. and Solberg, T. (1996) Characterizing filter media mineral coatings. *Journal AWWA*, **88** (12), 62-73.

O'Connor, J.T. (1971) Iron and Manganese. In *Water Quality and Treatment - A Handbook of Public Water Supplies.* Chapter 11, pp 378-396, McGraw Hill Book Company, New York.

Petrusevski, B., Boere, J., Shahidullah, S.M., Sharma, S.K. and Schippers, J.C. (2000) Adsorbent based point-of-use system for arsenic removal in rural areas. *Proceedings of IWA Conference on Innovations in Classical and Conventional Water Treatment Processes (26-29 September 2000) Amsterdam, The Netherlands*, 5, 1-13.

Salvato, J.A. (1992) *Environmental Engineering and Sanitation.* Fourth Edition, John Wiley & Sons, Inc., New York.

Stahl, R.S. and James, B.R. (1991) Zinc sorption by iron oxide coated sand as function of pH. *Soil Science Society of America Journal*, **55**, 1287-1290.

Standard Methods for the Examination of Water and Wastewater (1995) 19th edition, American Public Health Association/American Water Works Association/Water Environment Federation, Washington DC.

Stenkamp, V.S. and Benjamin, M.M. (1994) Effect of iron oxide coatings on sand filtration. *Journal AWWA*, **86** (8), 37-50.

Stumm, W. and Morgan, J.J. (1996) *Aquatic Chemistry - Chemical Equilibria and rates in natural waters.* Third Edition, John Wiley & Sons Inc., USA.

Takai, T. (1973) Studies on the mechanisms of catalytic deferrization (II) - Determination of chemical structure of the deferrizing catalyst. *Journal of Japan Water Works Association*, No 466, July, 22-33.

Tamura, H., Goto, K. and Nagayama, M. (1976) The effect of ferric hydroxide on the oxygenation of ferrous ions in neutral solutions. *Corrosion Science*, **16**, 197-207.

Twort, A.C., Ratnayaka, D.D. and Brandt, M. J. (2000) *Water Supply.* Fifth edition, Arnold Publishing - Hodder Headline Group, London, UK.

Viraraghavan, T. and Li, C. (1999) Removal of selenium from drinking water. *Water Supply*, **17** (3-4), 163-167.

Zhang, Y.H. (1997) Filtration media rids iron from groundwater. *Water/Engineering & Management*, December, 30-31.

Chapter 6

COMPARISON OF PHYSICOCHEMICAL IRON REMOVAL MECHANISMS IN FILTERS

Part of this Chapter was published by S.K.Sharma, J. Kappelhof, M. Groenendijk and J.C. Schippers (2001) *Journal of Water Supply: Research and Technology - Aqua*, **50** (4), 187-198.

ABSTRACT

Water supply companies are continually seeking means to improve the process efficiency of iron removal from groundwater in order to minimise the deposition of iron in distribution networks, the backwash water use, and the volume of sludge produced. It is expected that the process efficiency can be optimised with a better understanding of the different mechanisms involved in the iron removal process and their optimal utilisation. In order to investigate the options to improve the performance of an iron removal plant, pilot experiments were conducted with filters operating in two different iron removal modes, namely floc filtration and adsorptive filtration. The results were then compared with the performance of the full-scale groundwater treatment plant Gilze, which operates mainly in the floc filtration mode. Results demonstrated that the performance of WTP Gilze in terms of ripening time, filter run time and filtrate quality (residual iron and turbidity) could be improved by operating the filters a) mainly in adsorptive mode with fine sand (0.5-0.8 mm) and low influent oxygen, or b) mainly in floc filtration mode with dual media (anthracite 0.8-1.2 mm and sand 0.5-0.8 mm). In general, for a single media fine sand filter, the adsorptive mode gave a longer run time than the floc filtration mode with comparable filtrate quality. To facilitate adsorptive mode, oxygen content and pre-oxidation time should be kept as low as possible. However, to ensure sufficient removal of manganese and ammonium present in groundwater, a minimum oxygen concentration is required (about 2 mg/l in the case of WTP Gilze). Experimental results suggested that filters with coarser single media or dual media operating mainly under adsorptive mode with short pre-oxidation time could further improve filter performance and likely result in significant cost savings.

6.1 INTRODUCTION

Iron in water supplies causes various aesthetic and operational problems including bad taste, discoloration, staining and deposition in the distribution system leading to after growth and incidence of high turbidity (O'Connor 1971; Vigneswaran and Visvanathan 1995). The EC directive recommends that the iron concentration in drinking water should be ≤ 0.2 mg/l (EC 1998). In the Netherlands, the guideline value for iron in drinking water is ≤ 0.05 mg/l (VEWIN 1993) and several water supply companies are aiming at iron levels below 0.03 mg/l in order to minimise the distribution network maintenance costs. Another important consideration is the minimisation of the treatment costs. This can be achieved by lowering the percentage of backwash water used (e.g. in the Netherlands groundwater abstraction is charged with a tax of about 0.18 Euro/m^3) and by reducing the volume of sludge produced. In this context, several water supply companies are continually seeking the means to improve the efficiency of existing groundwater treatment systems.

Aeration followed by rapid sand filtration is the most commonly used method for the removal of iron from groundwater (O'Connor 1971; Salvato 1992). Different mechanisms (physical, chemical, and biological) may contribute to iron removal in filters and the mechanisms playing a predominant role depend on the groundwater quality and process conditions (Lerk 1965; Hatva 1988; Mouchet 1992). A better understanding of the different mechanisms involved in the iron removal process is therefore important to optimise the design and operation of iron removal filters in terms of run time, filtrate quality and overall treatment costs.

6.2 PHYSICOCHEMICAL IRON REMOVAL MECHANISMS

There are two main physicochemical mechanisms of iron removal in filters, namely: the oxidation-floc formation mechanism (floc filtration) and the adsorption-oxidation mechanism (adsorptive filtration/catalytic iron removal), as illustrated in Fig. 6.1.

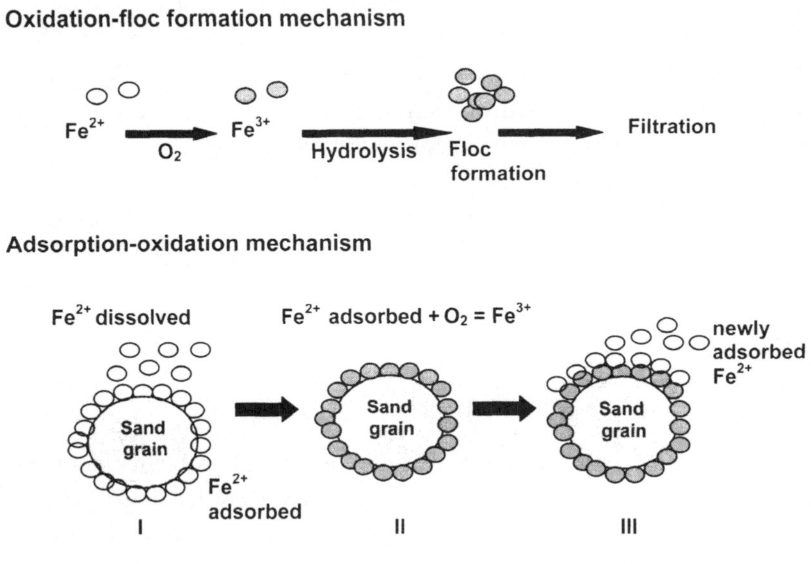

Fig. 6.1 Physicochemical iron removal mechanisms

In anaerobic groundwater, iron is commonly present in soluble iron(II) form. In the floc filtration mode of iron removal, commonly applied in many groundwater treatment plants, soluble iron(II) is first oxidized to insoluble iron(III) by aeration or chemical oxidation and the flocs formed are subsequently removed in a rapid sand filter (O'Connor 1971). In several conventional plants, however, oxidation is just partially achieved before filtration and as a consequence, besides oxidation-floc formation, adsorption-oxidation plays a role as well. In addition to rapid head loss development and related short filter runs, various other practical problems associated with conventional iron removal systems like long ripening periods, iron passing through the filter, and large volume of sludge produced have been reported (Mouchet 1992).

In adsorptive filtration, iron is mainly removed in the iron(II) form by adsorption onto the surface of the filter media. Adsorbed iron(II) is subsequently oxidised in the presence of oxygen to form an iron oxide coating on the filter media surface and hence the process continues. Adsorption-oxidation is one of the mechanisms responsible for removal of iron entering the filter bed in iron(II) form in conventional iron removal plants (Ghosh *et al.* 1967; Committee Report 1987) and is also the dominant mechanism in sub-surface iron removal (Rott 1985; Braester and Martinell 1988). For successful application of this mechanism, the design and operation of the process should ensure that the pre-oxidation of iron(II) is limited. When oxygen is present in the feed water, the oxidation of iron(II) during filtration cannot be avoided completely, therefore some flocs are formed and floc filtration inevitably plays a role. However, the extent of floc formation can be minimised by adapting the process conditions. It is expected that the adsorption-oxidation mechanism will result in lower head loss in comparison to the oxidation-floc formation mechanism, as adsorbed iron forms a compact coating rather than a voluminous floc. Thus, the filter runs are expected to be longer and consequently the backwash water requirement and the volume of sludge produced are reduced.

Adsorption of iron(II) onto the surface of filter media or onto iron hydroxide flocs is a well-known process (Cox 1964; Hult 1973; Tamura *et al.* 1976; Chapter 2). Adsorption based processes for heavy metal removal are receiving more attention these days as they have advantages over the conventional process in that they have high removal efficiencies and produce less or no sludge (Edwards and Benjamin 1989; Benjamin *et al.* 1996). However, no information is available on the effectiveness of the adsorption-oxidation mechanism when employed as the dominant mechanism of iron removal from groundwater in filters. In this study the performance of two filters, one operating predominantly in floc filtration mode and another predominantly in adsorptive mode, was compared at a pilot scale. The results of these experiments were compared with the performance of the full-scale ground water treatment plant (WTP) Gilze of N.V. Waterleiding Maatschappij "Noord-West-Brabant" (WNWB), in southern Netherlands.

In the WTP Gilze, the filtrate turbidity and iron concentrations after backwashing remain relatively high during the first 1.5 to 2 hours (ripening period) (Fig. 6.2). A similar problem is experienced after increasing the production rate (e.g. residual iron increases from 0.03 to 0.08 mg/l for more than 30 minutes when the filtration rate increases from 10 to 15 m/h) (Fig. 6.3). Additionally, the filtrate turbidity is relatively high (~ 0.3 NTU). The aim of this study is to explore the possibilities for more efficient iron removal in groundwater treatment plants and specifically to investigate the options for improving the performance of WTP Gilze. Therefore, pilot plant experiments were conducted to compare the performance of filters operating under two main physicochemical mechanisms of iron removal.

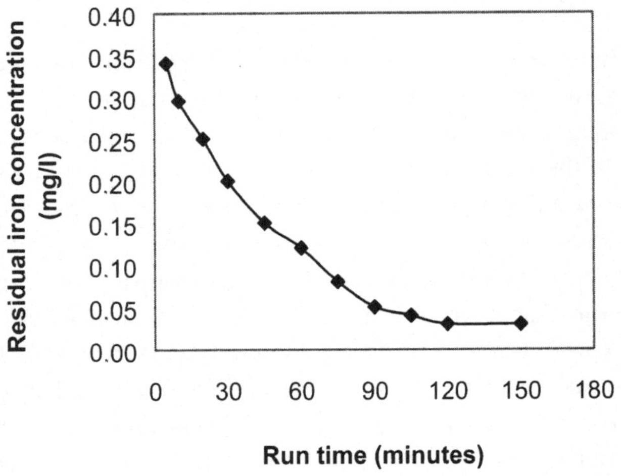

Fig. 6.2 Typical ripening curve of the full-scale filter at WTP Gilze after backwashing (filtration rate ~8 m/h)

6.3 MATERIALS AND METHODS

6.3.1 Water treatment plant Gilze

Gilze is one of the smaller WTP of WNWB with a production capacity of 2 million cubic meters per year. WTP Gilze uses groundwater as its source and employs the conventional treatment scheme: cascade aeration followed by rapid sand filtration. At WTP Gilze, 70% to 85% of iron is oxidized before entering the filter bed; hence oxidation-floc formation is the dominant iron removal mechanism. There are six sand (0.7-1.25 mm) filters with a total surface area of 37.8 m^2 and the filtration rates vary from 8 to 15 m/h depending upon the production capacity (0-150-225-375-435 m^3/h). A typical daily production pattern (pumping rate) at WTP Gilze is presented in Fig. 6.3. The depth of the supernatant water is 1.0 m and the bed depth is 1.8 m. The filters are backwashed after 48 to 72 hours at a backwash rate of 35-40 m/h as follows: water (7 min), air (5 min) and water (7 min).

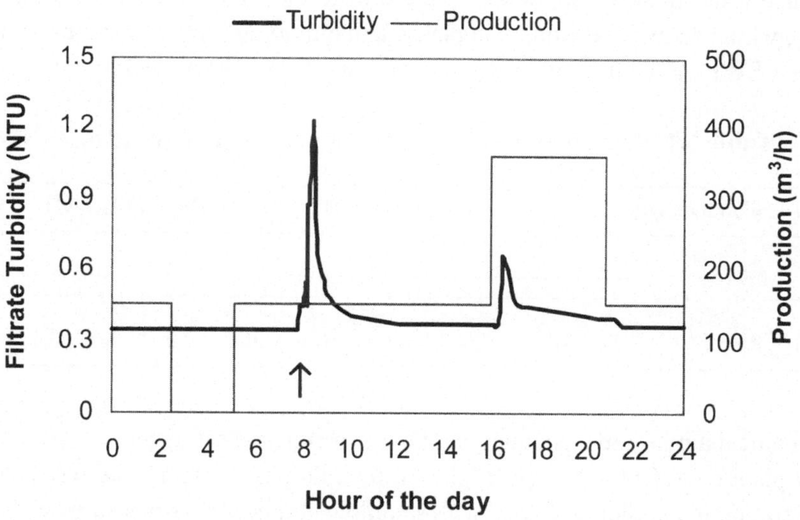

Fig. 6.3 Increase in filtrate turbidity with backwashing of filters and change in production rate at WTP Gilze (↑ Three filters are backwashed subsequently around 8:00 am)

6.3.2 Experimental set-up and process conditions

In this study, the experiments were carried out in two phases.

In the first phase, two pilot filters were equipped with fine sand media (0.5-0.8 mm) and operated in two different modes; one predominantly in floc filtration mode and another predominantly in adsorptive mode with different feed water oxygen concentrations (0 mg/l, 0.5-0.7 mg/l and 1-2 mg/l). These options were compared with the existing full-scale filter equipped with coarser single media (0.7-1.25 mm), with respect to ripening time, filter run time, filtrate quality and production efficiency. The minimum acceptable run time of 48 hours at the average filtration rate of 10 m/h was set by WNWB, based on operational considerations. The maximum allowable head loss in the filter bed was limited to 2.6 m and the guideline values for the filtrate quality, based on the VEWIN (Association of Water Works in the Netherlands) recommendations, were iron $\leq$ 0.05 mg/l, manganese $\leq$ 0.02 mg/l and ammonium $\leq$ 0.04 mg/l (VEWIN 1993). The WNWB internal guideline for iron concentration in drinking water entering a distribution system is $\leq$ 0.03 mg/l.

During the second phase, in an attempt to improve the performance of the filter operating predominantly under floc filtration mode, filter runs were conducted with the dual media (anthracite 0.8-1.2 mm and sand 0.5-0.8 mm) filter.

The average water quality parameters of the raw groundwater at WTP Gilze are presented in Table 6.1, which shows that iron, manganese and ammonium concentrations in raw water are above the VEWIN guideline values for drinking water (VEWIN 1993).

Table 6.1 Raw groundwater quality parameters at WTP Gilze, WNWB

Temp	pH	Conductivity	O_2	Fe	Mn	CO_2	HCO_3	Ca	Mg	SO_4	NH_4	Cl	TOC
°C		ms/m	----------------------------------mg/l--										
12	7.6	25.5	0.0	1.8	0.08	6.0	160	50	4.5	8	0.12	12	1.26

The schematic diagram and a photograph of the experimental set-up are shown in Figs. 6.4 (a-b). The pilot plant consisted of two plexi glass filter columns (300 mm internal diameter and 3 m high), operating in parallel, provided with sampling and head loss measurement taps. Columns were equipped with an automatic water level controller to maintain the constant supernatant level and filtration rate. Each column was equipped with a flow meter and a water meter to measure the rate of flow and the total production.

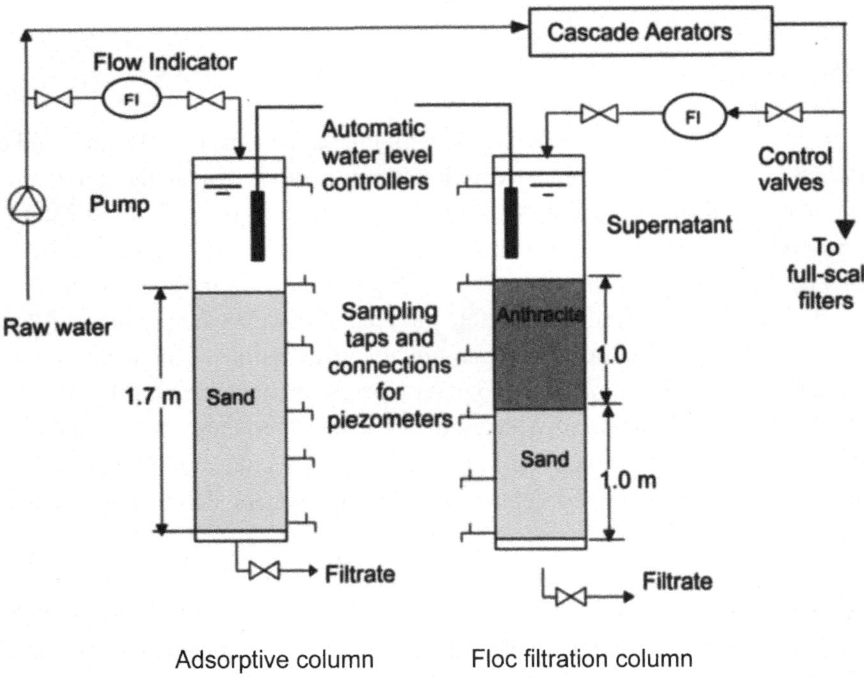

Fig. 6.4a Schematic diagram of the pilot plant experimental set-up

Fig. 6.4b Photograph of the pilot plant experimental set-up (floc filtration column)

The details on the filter media used and process conditions applied are included in Table 6.2.

Table 6.2 Experimental conditions

	Column 1	Column 2
Dominant mechanism	Oxidation-floc formation	Adsorption-oxidation
Filter media	Phase I - 1.8 m sand (0.5-0.8 mm) Phase II - 1.0 m sand (0.5-0.8 mm) + 1.0 m anthracite (0.8-1.2 mm)	1.7 m sand (0.5-0.8 mm)
Filtration rate	10m/h, 15m/h	10 m/h
Feed water	Aerated Gilze well water	i) Anaerobic Gilze well water ii) Gilze water with 1-2 mg/l O_2 iii) Gilze water with 0.5-0.7 mg/l O_2
Depth of supernatant	0.9-1.0 m	0.9-1.0 m

6.3.3 Experimental procedure

The filters were operated under different conditions to facilitate two different mechanisms of iron removal. Influent to the "floc filtration column" was the aerated water after the cascades of WTP Gilze so that most of the iron entered the filter bed in iron(III) form. Raw water directly from the wells was supplied to the "adsorptive column" without aeration to ensure that the majority of the iron was in iron(II) form. The pilot filters were operated at filtration rates of 10 and 15 m/h. The run time of the filters was determined by either filtrate quality breakthrough (iron concentration > 0.05 mg/l) or terminal head loss (> 2.6 m). At the end of the filter run the columns were backwashed and the filter runs were started again.

Backwashing: When operated under anoxic conditions, the adsorptive column was backwashed only with water (45 m/h) for 15 minutes. In the filter runs when the adsorptive column was operated with some oxygen in the influent, the backwash procedure was as follows: i) water (45 m/h) for 6 minutes, ii) air (50 m/h) + water (10 m/h) for 4 minutes, and iii) water (45 m/h) for 6 minutes. The dual media floc filtration column was backwashed as follows: i) water (45 m/h) for 6 minutes, ii) air (50 m/h) for 4 minutes, and iii) water (45 m/h) for 6 minutes.

Measurements: Head losses along the filter columns were measured periodically with piezometers. The pH and the dissolved oxygen concentrations of influent and filtrate turbidity were monitored on-line continuously. Iron concentrations in the influent, filtrate, and along the filter column were measured at different times during the filter runs. Iron(II), total iron, manganese, and ammonium contents of the periodic samples were measured immediately at the site with a Dr. Lange ISIS 9000 MDA photometer using LCK 320, LCW 032, and LCK 302 cuvettes (Operation Manual for Dr. Lange Photometer). Duplicate samples were taken frequently and concentrations of iron and manganese were also measured in the laboratory with a spectrophotometer (Perkin Elmer 550 S) and an atomic absorption spectrometer (Perkin Elmer AAS 3110) to verify the results of field measurements (Standard methods 1995). It was found that field measurements were comparable with the laboratory results.

6.4 RESULTS

6.4.1 Single media filter in floc filtration mode

Preliminary experiments were conducted with a single media sand (0.5-0.8 mm) filter operating at a filtration rate of 10 m/h in floc filtration mode. Under applied conditions, about 75% of the iron entered the filter bed in iron(III) form. The residual iron concentrations were <0.05 mg/l from the beginning of the filter run. The filter run time was, however, limited to only 22 to 24 hours due to excessive head loss (2.4 m in 24 hours). The full-scale filters at WTP Gilze have run times of 48 to 72 hours at a filtration rate of 10 to 15 m/h. This shows that sand media of 0.5–0.8 mm under floc filtration mode is not an acceptable option due to short run time; hence, this option was not investigated further.

6.4.2 Single media filter in adsorptive mode

Filter runs with anaerobic well water

Development of the iron oxide coating: Pilot plant experiments were started with new virgin sand. Under applied conditions, about 95% of the iron entered the filter bed in iron(II) form. At the end of each run the filter was backwashed to regenerate the adsorption sites and to remove any flocs which had formed. During the initial filter runs, the iron removal was poor (filtrate iron concentration >1 mg/l). The filtrate quality, however, improved considerably with the development of iron oxide coating on the filter media as the iron(II) adsorption capacity of the media increased with the coating development (Fig. 6.5). It took several weeks to develop an effective iron oxide coating and to achieve a good performance under the adsorptive filtration mode. After about 75 days of operation, the residual iron concentrations were consistently <0.05 mg/l.

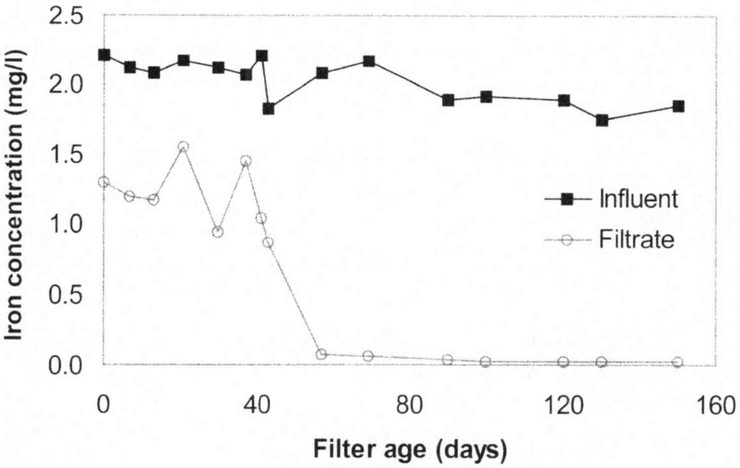

Fig. 6.5 Improvement in filtrate quality of the adsorptive filter operating under anaerobic conditions with filter age (filtration rate = 10 m/h, depth of supernatant = 1.0 m, samples taken one hour after backwashing)

That a rather longer time was required for the development of the coating is probably due to the operation of the adsorptive column under anoxic conditions, in which the oxidation of adsorbed iron and regeneration of the adsorption sites takes place during backwashing only. The development of the coating is likely to be faster if the feed water contained a small amount of oxygen, which would speed up the process of the coating development with the continuous oxidation of adsorbed iron and consequent regeneration of the adsorption sites. The conditions governing the development of the catalytic iron oxide coating, however, need to be investigated further.

Ripening after backwashing: After the coating development, the ripening time after backwashing was shortened to ≤ 30 minutes and the residual iron concentrations were consistently <0.05 mg/l. This is much faster than in the full-scale filter where filter ripening takes about 1.5-2 hours.

Filtrate quality and run time: The first few runs, after the development of the coating, produced filtrate of good quality (iron <0.05 mg/l, turbidity <0.1 NTU) with the run lengths ranging from 75 to 80 hours at a filtration rate of 10 m/h. However, further filter runs were inconsistent with respect to filtrate quality. During the later stage, filter run time varied between 24 to 30 hours. At WTP Gilze, there are regular night supply shutdowns of variable duration (Fig. 6.3). This might have introduced oxygen into the filter, which most likely formed the flocs through out the filter bed and regenerated the catalytic iron oxide coating occasionally by oxidizing the adsorbed iron and consequently resulted in the variable filter run times. The conditions governing the adsorption capacity and regeneration of the adsorption sites need further detailed investigation.

In all filter runs, the head loss development in the adsorptive column was very slow compared to the floc filtration column (Fig. 6.6) and the filter runs were terminated due to residual iron breakthrough. Under applied conditions, there was no manganese and ammonium removal.

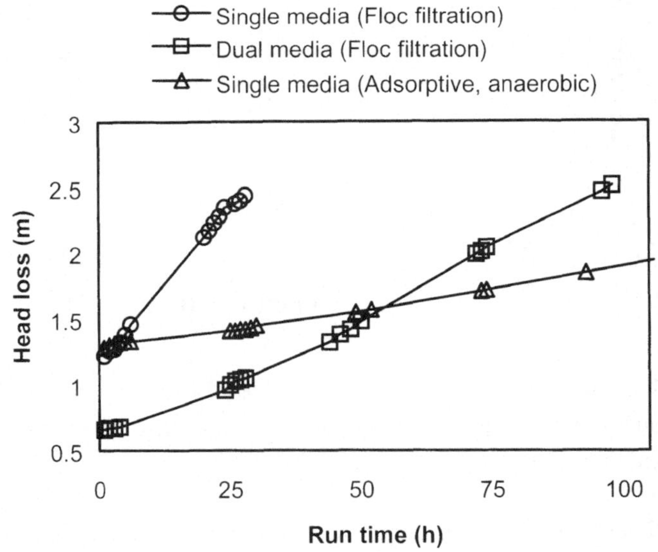

Fig. 6.6 Head loss development under different operational conditions (filtration rate = 10 m/h and depth of supernatant = 1.0 m)

Filter runs with 1-2 mg/l oxygen in influent

It was hypothesised that continuous regeneration of adsorption sites with the introduction of some oxygen in the feed water would increase the run time of the filter. Therefore, oxygen-rich treated water from WTP Gilze was mixed with anaerobic well water to get feed water with an oxygen concentration of 1-2 mg/l. This water was introduced at the top of the filter column; this resulted in a pre-oxidation time of about 8 minutes. Under these conditions, about 10-15% of the iron entered the filter bed in iron(III) form.

With this modification, the ripening time after backwashing was 45-60 minutes and the average filter run time was 44 hours. The filter runs, in terms of filtrate quality and filter run time, were consistent and reproducible. Residual iron and turbidity were consistently <0.05 mg/l and <0.1 NTU respectively, with an additional advantage of ammonium and manganese removal (from 0.15 to 0.06 mg/l and from 0.08 to 0.04 mg/l respectively). Filter runs were terminated due to excessive head loss (2.6 m in 44 hours). From the Stumm and Lee iron oxidation rate equation it was calculated that, under the applied conditions and oxidation rate constant $k = 1.14x\ 10^{13}$ $l^2/(mol^2\ atm.min)$, 16-29% of the dissolved iron entering the filter bed is oxidised within the filter (Stumm and Lee 1961). Therefore, the introduction of oxygen coupled with a relatively long pre-oxidation time in the supernatant, very likely resulted in substantial floc formation in the supernatant and in the filter bed, thereby causing the rapid head loss development and shortening filter runs.

Filter runs with 0.5-0.7 mg/l oxygen in influent

To further prolong filter run length by reducing head loss, the adsorptive column was operated with an influent oxygen concentration of about 0.6 ± 0.1 mg/l and feed water was introduced 10 cm above the surface of the filter media to shorten the pre-oxidation time to 1 minute. Under these conditions, the amount of iron entering the filter bed in iron(III) form was reduced to about 8-10%.

Under these conditions, the ripening time after backwashing was 30-60 minutes and the average filter run time further increased from 44 hours to above 50 hours. At a filtration rate of 10 m/h, after filter ripening, residual iron concentration and turbidity were consistently <0.05 mg/l and <0.1 NTU respectively. Ammonium removal was rather good (from 0.15 to 0.01 mg/l) but manganese removal was rather poor (from 0.08 to 0.07 mg/l). Head loss development was still limiting the filter run length to 50-72 hours. Calculations using the Stumm and Lee equation show that under the applied operating conditions only 8-12% of the iron entering the filter bed is oxidised within the filter (Stumm and Lee 1961). This explains why the head loss developments in these runs were much slower than in the filter runs with 1-2 mg/l of oxygen.

Head loss development was substantial in the first 0.4 m of the filter bed (Fig. 6.7), presumably due to the deposition of flocs (Fig. 6.8) and thereafter in the deeper layers further head loss increase was very marginal. It is clear from the Lindquist diagram (Fig. 6.7) that filter runs much longer than 50 hours are possible without developing negative pressure.

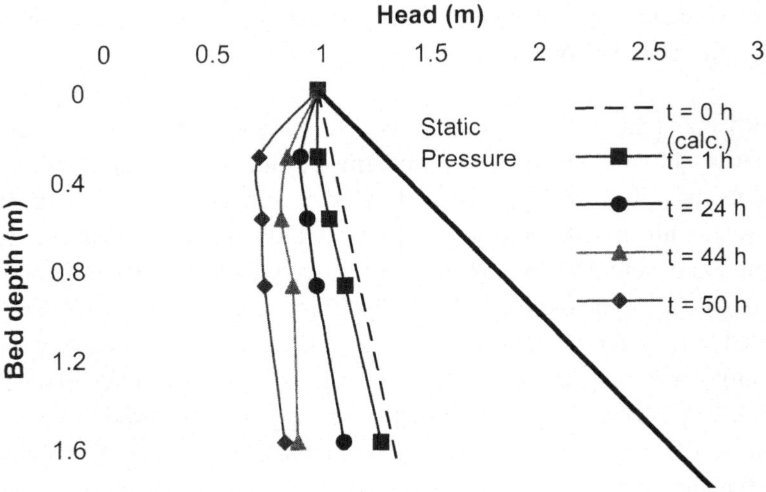

Fig. 6.7 Lindquist diagram of adsorptive filter with feed water oxygen of 0.5-0.7 mg/l (filtration rate = 10 m/h, depth of supernatant = 0.94 m)

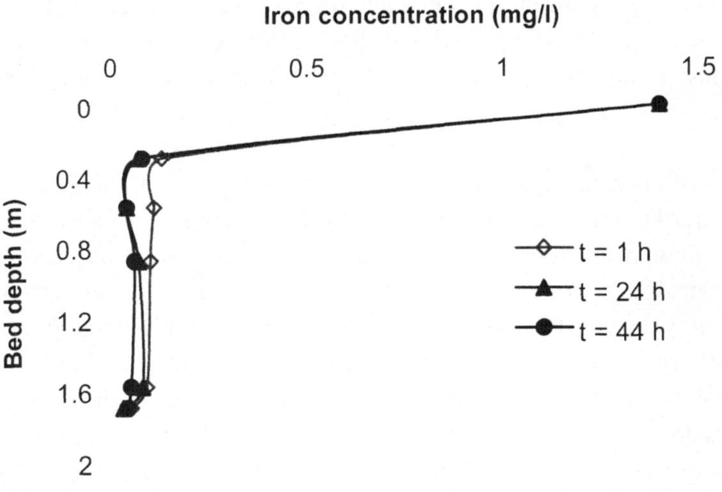

Fig. 6.8 Iron removal profile for adsorptive filter with 0.5-0.7 mg/l of oxygen in feed water (filtration rate = 10 m/h, depth of supernatant = 0.94 m)

Table 6.3 summarises the established evaluation criteria and results obtained with pilot filters equipped with single sand media (0.5-0.8 mm) operating under different process conditions and compares them with the results obtained with the full scale filter. For all the options examined, the ripening time was shortened to ≤ 60 minutes. It is evident from Table 6.3 that, under given conditions, the adsorption–oxidation mechanism is a better option than the oxidation-floc formation mechanism. The adsorption-oxidation mechanism results in longer filter runs, lower head loss development and lower sludge volume, therefore reducing the amount of backwash water used and sludge treatment and disposal costs. An estimation of operation and maintenance cost savings with adsorptive filtration, under the given conditions, is presented in Appendix 6.1. A single media fine sand filter operating under adsorptive mode with 0.5-0.7 mg/l of oxygen in the influent satisfies WNWB filter run time criterion. However, in pilot filters operated under adsorptive mode, manganese and ammonium were not sufficiently removed.

Table 6.3 Overview of comparison criteria and results for the single media filters

Parameter	Reference Criteria	Full-scale filter (Existing)	Single media fine sand filter (Pilot)			
Dominant removal Mechanism		Oxidation-floc formation	Oxidation-floc formation	Adsorption - oxidation		
				Anaerobic	0.5-0.7 mg/l oxygen	1-2 mg/l oxygen
Ripening time (minute)	≤60	90 - 120	<30	<30	30 - 60	45 - 60
Run time (h)	>48	48 - 72	22 – 24	24 - 30	50 - 72	40 - 44
Head loss (m)	<2.6		2.2 – 2.4	1.4 - 1.5	1.7 - 2.5	2.2 - 2.6
Run time Criteria		Fixed time	Head loss	Filtrate quality	Head loss	Head loss
Filtrate quality						
Fe (mg/l)[a]	<0.05	0.03	0.03	0.03	0.04	0.02
Mn (mg/l)[a]	<0.02	0.02	NM	0.07	0.07	0.04
NH_4^+ (mg/l)[a]	<0.04	0.05	NM	0.15	0.01	0.06
Production efficiency (%)[b]	≥98	97	96	96	98	98

a. Average concentrations
b. Calculated as difference between total filtered water (100%) and % of water used for backwashing
NM = not measured

Analysis of head loss development in filters

The initial head loss H_0 in the clean filter with laminar flow (Re <2) is given by Carman–Kozeny equation,

$$H_0 = \frac{180\,\nu\,(1-\varepsilon)^2\,v}{g\,\varepsilon^3\,d^2}L \qquad (6.1)$$

where Re = Reynolds number = v d/< (-)
< = kinematic viscosity of water (m^2/s)
ε = clean bed porosity (-)
v = filtration rate (m/s)
L = bed depth of filter media (m)
g = acceleration due to gravity = 9.8 m/s^2
d = specific diameter (m)

The theoretical head loss development in the single media fine sand (0.5-0.8 mm) filter operating in adsorptive filtration mode under anoxic condition was calculated from equation (6.1), assuming that (i) all iron is removed at the top 30 cm of the filter bed, (ii) the filter bed porosity decreases with the adsorption of iron(II) onto the surface of the filter media, and (iii) the density of the iron oxide coating is 2200 kg/m^3. The calculated head loss development in the "purely" anoxic adsorptive filter was then compared with the head loss developments observed experimentally with different oxygen concentrations in the feed water (Fig. 6.9).

Figure 6.9 shows that the oxygen concentration in the feed water had a strong influence on the head loss development, which decreased with decreasing oxygen concentration. The head loss development was fastest in the filter operated under floc filtration mode with an average of 8.5 mg/l of oxygen in the feed water and 75% of the iron entering the filter bed in iron(III) form. In the case of adsorptive filters, feed water with 1-2 mg/l of oxygen and 10-15% of iron entering the filter bed as iron(III) gave a much faster rate of head loss development compared to feed water with 0.5-0.7 mg/l of oxygen and 8-10% of iron entering the filter bed as iron(III).

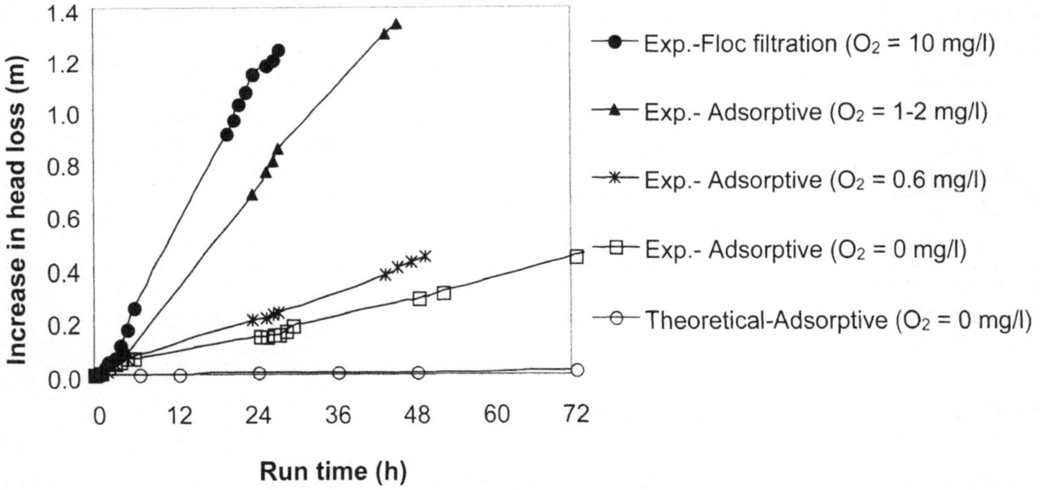

Fig. 6.9 Increase in head loss development in filters with different oxygen concentrations in the feed water (filtration rate = 10 m/h, grain size = 0.5-0.8 mm, bed depth = 1.7 m)

The head loss development was least with the unaerated raw ground water of WTP Gilze, however this was still higher than the theoretically possible head loss development in an adsorptive filter operating under anoxic condition. Theoretical head loss calculations showed that the head loss development as observed in the case of the experimental adsorptive filter operating under anoxic condition would occur only when all the iron is removed in the top 3 mm of the filter bed, which is most unlikely the case. This implies that, also in the case of adsorptive filter operating under anoxic condition, there was probably some iron(II) oxidation and floc formation due to the presence of a small amount of oxygen.

The rate of head loss development (slopes of the head loss curves) under different conditions are summarised in Table 6.4.

Table 6.4 Rates of head loss development under different conditions

Dominant mode of operation	Average oxygen concentration (mg/l)	Rate of head loss development (mm/h)
Adsorptive filtration (theoretical)	0	0.2
Adsorptive filtration	~ 0	5.2
Adsorptive filtration	0.6	8.8
Adsorptive filtration	1.5	28.9
Floc Filtration	8.5	46.0

An interpolation of the rate of head loss development and the oxygen concentrations in the feed water (Fig. 6.9) gave that in the case of the experimental adsorptive filter operated under anoxic conditions, the average oxygen concentration in the feed water should have been about 0.3 mg/l instead of 0 mg/l. Intrusion of oxygen in the system can not be excluded, because it is well known that oxygen can enter the raw water e.g. via wells and valves. Iron(II) oxidation and floc formation due to this 0.3 mg/l of oxygen in the feed water might be responsible for the observed head loss development.

6.4.3 Dual media filter in floc filtration mode

Additional filter runs were conducted with the pilot filter equipped with dual media (anthracite and sand) in order to reduce the head loss development and increase the run time. Under applied conditions, about 75% of the iron entered the filter bed in iron(III) form.

Ripening after backwashing: Under these conditions, the ripening time was reduced to about 30-60 minutes. Figure 6.10 clearly shows that the ripening of the dual media filter is much faster than that of the full-scale filter, likely due to the polishing effect of the fine sand layer.

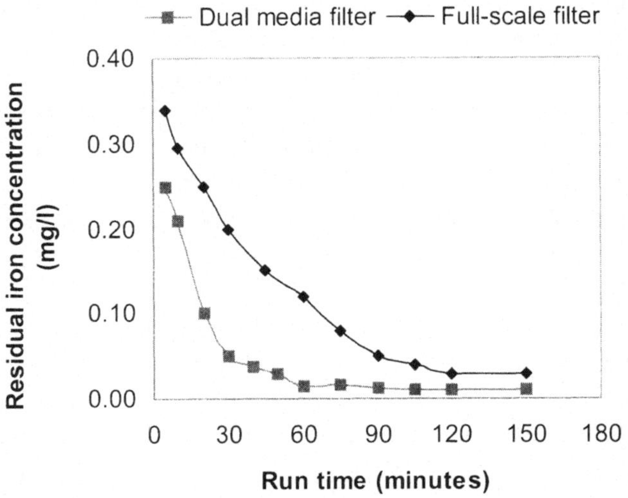

Fig. 6.10 Ripening of the dual media filter in floc filtration mode and full-scale filter after backwashing (filtration rates: Dual media filter = 10 m/h, Full-scale filter ~8 m/h)

Filtrate quality and runtime: The residual iron concentrations and turbidity after ripening were <0.05 mg/l and <0.1 NTU respectively throughout the filter run. The filter run time was determined by the head loss development. At a filtration rate of 10 m/h, the filter runs were typically 85-90 hours, with a production efficiency of about 99%. A typical Lindquist diagram (Fig. 6.11) shows that most of the head loss is developed in the anthracite layer. It was also observed that, in this case, most of the iron removal occurs in the anthracite layer while the sand bed serves as a polishing step. For this option, the residual manganese and ammonium concentrations were also well below the guideline values (VEWIN 1993).

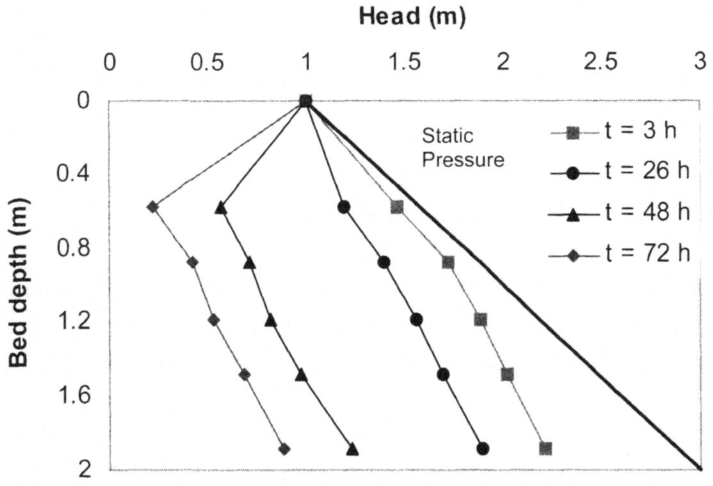

Fig. 6.11 Typical Lindquist diagram of the dual media filter operating under floc filtration mode (filtration rate = 10 m/h, depth of supernatant = 1.0 m)

Effect of changing production rate: An increase in the filtration rate from 10 m/h to 15 m/h during the filter run resulted only in a slight temporary increase in residual iron concentration (from 0.02 mg/l to 0.04 mg/l) for about 20 minutes. A similar increase of the filtration rate in the existing full-scale filter resulted in an increase of residual iron from 0.03 to 0.08 mg/l that lasted for more than 30 minutes.

These results show that use of dual media improved the performance of the filter operating under floc filtration mode and satisfied all the evaluation criteria. Hence, a dual media filter operating under floc filtration mode might be an attractive alternative for WTP Gilze.

6.4.4 Quantification of iron passing the filters

The average concentrations of iron passing through filters operating under different conditions during the entire filter run were quantified by computing the area under the residual iron concentration versus time curve (Table 6.5). The dual media filter operating in floc filtration mode and the adsorptive filter with influent oxygen of 1–2 mg/l result in the lowest amount of iron passing the filter. Operation of the full-scale filters under these conditions could consequently provide savings in terms of operation and maintenance of the distribution network as well as reduce "red water" incidents. The full-scale filter and the adsorptive filter with influent oxygen of 0.5-0.7 mg/l also satisfy the VEWIN guideline value for iron of 0.05 mg/l but slightly exceed the WNWB guideline for iron concentration in drinking water entering a distribution system (<0.03 mg/l).

Table 6.5 Iron passing through the filters (Filtration rate =10 m/h)

Filter	Iron Passing[a] mg/l
Full scale filter No.1[b]	0.03
Dual media filter in floc filtration mode	0.02
Single media adsorptive filter (0.5-0.7 mg/l O_2)	0.04
Single media adsorptive filter (1–2 mg/l O_2)	0.02

a. Average for a filter run *b. Filtration rate 8 m/h*

6.5 DISCUSSION

Experimental results strongly suggest that adsorptive iron removal is an attractive iron removal alternative for some groundwaters. The influent oxygen concentration has a strong influence on the removal efficiency of iron, manganese and ammonium. Theoretically, a minimum of 1.0 mg/l of oxygen is needed for the complete oxidation of iron, manganese and ammonium present in the groundwater of Gilze. Sufficient oxygen and nitrifying bacteria should be present for ammonium removal. Therefore, it is very likely that nitrification and manganese removal in the filter runs in adsorptive modes with 0 mg/l and 0.5-0.7 mg/l of influent oxygen were limited due to a lack of oxygen. Consequently, to improve manganese and ammonium removal in Gilze, it is advisable to operate an adsorptive filter with a feed water oxygen concentration of about 2 mg/l, assuming that filter design is optimised with respect to head loss development and filter run time. The required oxygen concentration in influent can be achieved by aerating only a fraction of the total raw water supply and mixing it with the remaining anoxic raw groundwater before filtration step. Alternatively, some oxygen-rich treated water e.g. finished/drinking water can be mixed with anoxic groundwater to control the oxygen concentration in the feed water. To reduce iron oxidation and subsequent floc formation, the feed water should be introduced just above the media surface, thus reducing pre-oxidation time to a minimum. In addition, the head loss development in the filter bed can be reduced and, consequently, the filter run time can be prolonged by facilitating deep bed filtration with the use of the dual media or the coarser single media as used in the full-scale filters.

The results obtained strongly suggest that a dual media filter operating mainly in adsorptive mode with a short pre-oxidation time is an attractive option worth further investigation for WTP Gilze. Another promising alternative is to run two filters in series at a high filtration rate of 20 m/h. The first filter equipped with coarser media operating in the adsorptive mode with very low oxygen concentration would remove the majority of the iron, and the second filter with finer media operating in the floc filtration mode could be optimised to remove manganese, ammonium, and the remaining iron. With this arrangement both filters can run at higher filtration rate, hence the overall process does not need more filtration surface area than the single step filtration scheme.

When methane is also present in groundwater, it can be removed by vacuum stripping before adsorptive iron removal. Alternatively, methane can also be removed by stripping after the adsorption of iron(II). However, for this option adsorptive filter should run under anaerobic conditions and regenerated intermittently. This is because under aerobic conditions methane is easily oxidised by bacteria forming larger quantities of biomass.

Presently there are no other methods except continuous iron measurement to know the saturation of the adsorptive filter running under intermittent regeneration mode. This is certainly a disadvantage for the filters operating under anoxic conditions. However, when the filters are running under aerobic conditions, the problem of saturation would not arise as the adsorption sites are continuously regenerated. Therefore, the criteria for backwashing in this case will be exceeding the maximum head loss allowed.

Rapid development of an effective iron oxide coating that will reduce "the ripening time of new media" is important for successful implementation of this mechanism. Secondly, in the adsorptive mode the filter media will grow with time. Hence, long-term experiments are required to determine the method of rapid development of an effective coating and to assess the rate of growth of media and the required frequency of the filter media replacement.

6.6 CONCLUSIONS

- The filter with fine sand (0.5-0.8 mm) operating in floc filtration mode gave better filtrate quality and a much shorter ripening time than the full-scale plant. However, filter run times were unacceptably short.
- The single media filter with fine sand running mainly in adsorptive mode produced filtrate with low iron concentrations and had short ripening times. However, manganese and ammonium removal were not satisfactory due to low oxygen concentration in the feed water. In general, for the single media fine sand filter, the adsorptive mode gave a longer run time than the floc filtration mode. The filter run times of the adsorptive filter were, however, influenced by the influent oxygen concentrations. Filter run time criteria established for WTP Gilze (48 hours) was easily fulfilled at an influent oxygen concentration of 0.5-0.7 mg/l, while at an influent oxygen concentration of 1-2 mg/l, the filter run times were slightly shorter (40-44 hours).
- The use of dual media (anthracite 0.8-1.2 mm and sand 0.5-0.8 mm) considerably prolonged the run time of the filter operating mainly in floc filtration mode and the filtrate quality satisfied all the evaluation criteria.
- Three options considered to be attractive to improve the performance of WTP Gilze are:
  - i) dual media filter in floc filtration mode.
  - ii) single media filter in adsorptive mode with coarser material (e.g. 0.7-1.2 mm) and low influent oxygen concentration (about 2 mg/l).
  - iii) dual media filter in adsorptive mode with low influent oxygen concentration (about 2 mg/l) and limited pre-oxidation time. This option needs to be verified in pilot or demonstration study.

  The experimental results suggest that the third option would result in the best performance in terms of both productivity and quality. However, for filters running under mainly adsorptive mode, the process conditions need to be selected carefully to ensure sufficient manganese and ammonium removal as well.

REFERENCES

Benjamin M.M., Sletten R.S., Bailey, R.P. and Bennett T. (1996) Sorption and filtration of metals using iron oxide coated sand. *Water Research*, **30** (11), 2609-2620.

Braester C. and Martinell R. (1988) The Vyredox and Nitredox methods of in situ treatment of groundwater. *Water Science and Technology*, **20** (3), 149-163.

Committee Report (1987) Research needs for the treatment of iron and manganese- Trace inorganic substance subcommittee. *Journal AWWA*, **79** (9), 119-122.

Cox, C.R. (1964) *Operation and Control of Water Treatment Processes.* WHO, Geneva, 209-218.

Edwards M. and Benjamin, M.M. (1989) Adsorptive filtration using coated sand a new approach for the treatment of metal bearing wastes. *Journal Water Pollution Control Federation*, **61** (9), 1523-1533.

EC (1998) *Official Journal of the European Communities* December 5 1998; Council Directive 98/83/EC on the quality of water intended for human consumption. L330/32 - L330/50.

Ghosh, M.M., O'Connor J.T. and Engelbrecht R.S. (1967) Removal of iron from groundwater by filtration. *Journal AWWA*, **59** (7), 878-896.

Hatva, T. (1988) Treatment of groundwater with slow sand filtration. *Water Science and Technology*, **20** (3), 141-147.

Hult, A. (1973) Filtration of iron during and after oxidation. *Effluent and Water Treatment Journal*, April, 209–215.

Lerk, C.F. (1965) *Enkele Aspecten Van De Ontijzering Van Grondwater* (in Dutch). PhD Dissertation, Technical University Delft, The Netherlands.

Mouchet, P. (1992) From conventional to biological removal of iron and manganese in France. *Journal AWWA*, **84** (4), 158-167.

O'Connor, J.T. (1971) Iron and Manganese. In: *Water Quality and Treatment - A Handbook of Public Water Supplies.* Chapter 11, pp. 378-396. McGraw Hill, New York.

Operation Manual for Dr. Lange ISIS 9000 MDA photometer and instruction leaflets for different cuvette tests, Dr Lange laboratories, Dusseldorf, Germany (DIN 38 405-D17)

Rott, U. (1985) Physical, chemical and biological aspects of the removal of iron and manganese underground. *Water Supply*, **3** (2), 143-150.

Salvato, J.A. (1992) *Environmental Engineering and Sanitation.* Fourth Edition, John Wiley and Sons, Inc. New York.

Standard Methods for the Examination of Water and Wastewater. (1995) 19th edition, American Public Health Association / American Water Works Association / Water Environment Federation, Washington, DC.

Stumm, W. and Lee G.F. (1961) Oxygenation of ferrous iron. *Industrial Engineering Chemistry*, **53** (2), 143-146.

Tamura H., Goto K. and Nagayama, M. (1976) The effect of ferric hydroxide on the oxygenation of ferrous ions in neutral solutions. *Corrosion Science*, **16,** 197-207.

VEWIN Recommendations 1993 (In Dutch: VEWIN Aanbevelingen 1993 - a. de hoedanigheid van het water = waterkwalitiet). VEWIN, The Netherlands.

Vigneswaran, S. and Visvanathan, C. (1995) *Water Treatment Processes - Simple Options.* CRC Press.

APPENDIX 6.1 Estimation of potential operation and maintenance cost savings with adsorptive iron removal

The following are the sample calculations of the operation and maintenance cost savings when a single media fine sand iron removal filter is operated in the adsorptive filtration mode instead of the oxidation-floc formation mode.

The calculations are based on the results of the pilot plant experiments at WTP Gilze, WNWB and the findings that the adsorptive filtration mode gives better filtrate quality and longer filter run than the oxidation-floc formation mode.

Assumptions:

Filter area	40	m^2
Filtration rate (average)	10	m/h
Production rate	400	m^3/h
Annual production	3,456,000	m^3
Annual production cost	2,350,080	Euro
Run time of the conventional filter	24	h
Run time of adsorptive filter	48	h

A. *Savings in backwash water*

Backwash water use	40	$m^3/h/m^2$
Duration	20	min
Total water savings per filter run	533	m^3
Annual water savings	96,000	m^3
Cost of treated water including groundwater tax and VAT	0.68	Euro/m^3
Cost savings	65,280	Euro
Total cost savings per m^3	0.019	Euro

B. *Savings in distribution system cleaning*

Average cost of distribution system cleaning per m^3 of water transported	0.007	Euro
Reduction in frequency of distribution system cleaning with adsorptive iron removal	50%	
Total cost savings per m^3	0.004	Euro

C. Savings in backwash water treatment and disposal

Reduction in backwash water volume per year	96,000	m^3
Cost of backwash water treatment	0.70	Euro/m^3
Cost savings	67,200	Euro
Total cost savings per m^3	0.019	Euro

D. Savings in operation and maintenance of aerators and energy costs

Total cost savings per m^3	0.005	Euro

Therefore, total operation and maintenance cost savings per m^3 of water produced (A+B+C+D)	0.047	Euro
Total annual cost savings	162,000	Euro (7%)

Similarly, it was calculated that when the filter run time of the adsorptive filter is 72 hours, the operation and maintenance cost savings will be 0.060 Euro/m^3, resulting in the total annual cost savings of 206,000 Euro (9%).

Reference for cost figures used:

Reijnen, G. (2001) *Personal communication.* Water Company Limburg (WML), The Netherlands.

Martijn Groenendijk (2001) *Personal Communication.* Water Company North West Brabant (WNWB), The Netherlands.

Chapter 7

MODELLING ADSORPTIVE IRON REMOVAL FROM GROUNDWATER

Part of this Chapter was published by S.K. Sharma, B.S. Mendis, M.R. Greetham and J.C. Schippers (2000) *Water Supply*, **18** (1-2), 604-608.

ABSTRACT

Iron removal filters with iron oxide coated sand are expected to perform better in terms of filtrate quality, run time and costs by changing the governing mode of operation from floc filtration to adsorptive filtration. Modelling of adsorptive iron removal is an essential tool to predict the performance of filters operated under adsorptive mode and to optimise design parameters for adsorptive iron removal filters. Of the two main steps in adsorptive iron removal, namely (i) adsorption of iron(II) onto filter media, and (ii) subsequent oxidation of adsorbed iron(II), this study focuses on the first step. The iron concentration in filtrates of adsorptive filters with new and iron oxide coated sand operating under anoxic conditions only, was modelled using adsorption isotherm, mass balance, and mass transfer equations. Kinetic rate constants, isotherm parameters, and mass transfer coefficients used for model predictions were determined from batch and short column experiments. The measured filtrate iron concentrations of laboratory filter columns were then compared with the predictions of three different fixed bed adsorption models. Consideration of external mass transfer alone (Constant Pattern Model, CPM) was not sufficient to predict the increase of filtrate iron concentration with time. Further improvement in prediction was observed when the Linear Driving Force Model (LDFM) was used and the effect of dispersion was also included in the model. The inclusion of surface diffusion (Plug Flow Homogeneous Surface Diffusion Model, PFHSDM) improved model predictions significantly than with the CPM. The LDFM and the PFHSDM predictions of iron breakthrough were better in the case of new sand as compared to iron oxide coated sand. The difference in model predictions and experimental results in the case of coated sand was likely due to initial pH drop in the pores of the filter media with iron(II) adsorption, and a consequent decrease in iron(II) adsorption capacity. Further improvement of the model prediction could be achieved by including the effect of pH change in pores of the filter media with iron(II) adsorption, inhomogeneity of the pore sizes, and the traces of the oxygen present in the water and in the pores of the filter media.

7.1 INTRODUCTION

Iron oxide coated filter media from iron removal plants have a high iron(II) adsorption capacity (see Chapter 2) which can be utilised to improve iron removal in filters. Existing iron removal filters with iron oxide coated sand are expected to perform better in terms of filtrate quality, filter run time and overall treatment costs by simply switching the mode of operation from floc filtration to adsorptive filtration.

A model describing adsorptive iron removal in filters could be an essential tool to predict the performance of existing filters when operated in adsorptive mode. Such a model could also be used to evaluate the effects of various process and operational variables on process efficiency and, consequently, help to establish optimal design parameters for iron removal filters operating in adsorptive mode. Additionally, prediction of the breakthrough with a mathematical model can reduce the need for lengthy and costly pilot experiments.

As discussed in Chapter 1 section 4.2, to achieve principally adsorptive iron removal, the filters can be operated in the following two modes: i) intermittent regeneration mode and ii) continuous regeneration mode. In both of these modes, two different steps can be identified, namely (i) adsorption of iron(II) onto filter media, and (ii) subsequent oxidation of adsorbed iron(II). This chapter deals with the modelling and prediction of iron(II) breakthrough in filtration columns with new and iron oxide coated sand under anoxic conditions and at different pH values.

7.2 THEORETICAL BACKGROUND

7.2.1 Reaction mechanism

Adsorption of iron(II) onto the surface of filter media can be represented by (Takai 1973; Barry *et al.* 1994)

$$\equiv \text{S-OH} + Fe^{2+} \leftrightarrows \equiv \text{S-OFe}^{+} + H^{+} \tag{7.1}$$

where ≡ S represents the surface of the filter media.

In the presence of oxygen, the adsorbed iron(II) is oxidized creating a new surface.

$$\equiv \text{S}-\text{OFe}^{+} + \frac{1}{4}O_2 + \frac{3}{2}H_2O \rightarrow \equiv \text{S}-\text{OFe}\langle^{\text{OH}}_{\text{OH}} + H^{+} \tag{7.2}$$

From (7.1) and (7.2)

$$\equiv \text{S}-\text{OH} + Fe^{2+} + \frac{1}{4}O_2 + \frac{3}{2}H_2O \rightarrow \equiv \text{S}-\text{OFe}\langle^{\text{OH}}_{\text{OH}} + 2H^{+} \tag{7.3}$$

Thus, the adsorption sites are regenerated and hence the process continues. The process is therefore sometimes referred to as '*Catalytic Iron Removal*'.

7.2.2 Prediction of breakthrough

The intermittent regeneration mode of adsorptive iron removal involves the following steps: (i) mass transfer of iron(II) from the bulk solution to the surface of the filter media, (ii) adsorption of iron(II) onto the outer surface of the filter media, (iii) diffusion of iron(II) into the internal pores followed by adsorption, (iv) exhaustion of the iron(II) adsorption capacity of the media and advancement of the adsorption front, (v) breakthrough of iron(II) in the filtrate, and (vi) regeneration of the adsorption surface with oxygen-rich water.

For this heterogeneous adsorption process, two main steps are required before adsorbates can be attached to the adsorbent surface: (i) transport of iron(II) from the bulk solution to the adsorbent surface (mass transfer resistance), and (ii) adsorption of the iron(II) on the adsorbent sites (surface reaction resistance). Adsorption reactions are generally very fast (Ruthven 1984; Tien 1994), therefore the mechanism of adsorption of iron(II) onto the surface of the filter media is assumed to be mass transfer controlled.

A fixed bed adsorptive iron removal process can be described by the macroscopic conservation equations, combined with the relevant adsorption equilibrium relationship, and initial and boundary conditions. The mass balances for an isothermal plug flow system with dispersion, consisting of spherical particles, can be written as (Ruthven 1984; Faust and Aly 1987; Tien 1994)

$$v_i \frac{\partial c}{\partial z} + \rho \frac{(1-\varepsilon)}{\varepsilon}\frac{\partial q}{\partial t} + \frac{\partial c}{\partial t} - D_L \frac{\partial^2 c}{\partial z^2} = 0 \qquad (7.4)$$

$$\frac{\partial q}{\partial t} = D_s \left(\frac{\partial^2 q}{\partial r^2} + \frac{2\partial q}{r\partial r} \right) \qquad (7.5)$$

Adsorption isotherm $q_s = f(c_s)$ (7.6)

The initial and boundary conditions are

$$c = c_0 + \left(\frac{D_L}{v_i} \right) \frac{\partial c}{\partial z} \quad at \quad z = 0 \qquad (7.7)$$

$$\frac{\partial c}{\partial z} = 0 \quad at \quad z = L \qquad (7.8)$$

$$\frac{\partial q}{\partial r} = 0 \quad at \quad r = 0 \qquad (7.9)$$

$$\frac{\partial q}{\partial r} = \frac{k_f}{D_s \rho}(c - c_s) \quad at \quad r = R \qquad (7.10)$$

$$c = 0 \quad at \quad t = 0 \qquad (7.11)$$

$$q = 0 \quad at \quad t = 0 \qquad (7.12)$$

where,

q = solid phase iron(II) concentration (kg/kg)
D_L = dispersion coefficient (m^2/s)
q_s = solid phase iron(II) concentration in equilibrium with c_s (kg/kg)
D_s = effective surface diffusivity (m^2/s)
r = radial distance in particle (m)
z = longitudinal distance in the column (m)

t = time (s)
c = liquid phase iron(II) concentration (kg/m^3)
c_s = liquid phase iron(II) concentration near the particle surface (kg/m^3)
c_0 = influent iron(II) concentration (kg/m^3)
v_i = interstitial liquid velocity (m/s)
k_f = external mass transfer coefficient (m/s)
R = radius of the filter bed particles (m)
ε = fixed bed porosity (dimensionless)
ρ = particle density (kg/m^3)
L = length of the column (m)

Several analytical and numerical solutions of these equations are available for the prediction of breakthrough in fixed bed adsorbers (Ruthven 1984; Hand *et al.* 1984; Sontheimer *et al.* 1988; Smith 1991; Tien 1994). The prediction of iron breakthrough requires an accurate determination of the isotherm parameters (relationship between q_s and c_s), the mass transfer coefficients k_f and D_s, and the dispersion coefficient D_L. In this study, the iron(II) breakthrough in the adsorptive filter was predicted by solving the above set of equations using three different models with increasing complexity, namely: Constant Pattern Model, Linear Driving Force Model, and Plug Flow Homogeneous Surface Diffusion Model. The predicted values were then compared with the results of laboratory-scale filter column experiments.

1. Constant Pattern Model

As the first simplification, it was hypothesised that the internal surface area of the filter media plays a minor role in the adsorptive iron removal process and there is no axial dispersion in the filtration columns ($D_L = 0$). Hence, the influence of internal mass transfer and axial dispersion on iron(II) adsorption was neglected. For this case, the iron(II) breakthrough was predicted using the constant pattern solution of an isothermal plug flow system with a linear-rate liquid-film mass transfer resistance model for Langmuir equilibrium given by Michaels (1952) for fixed bed ion exchange (Ruthven 1984; Wankat 1990). For this case the rate equation (7.5) can be replaced by

$$\frac{\partial q}{\partial t} = k'(c - c_s) = \frac{3k_f}{R\rho(1-\varepsilon)}(c - c_s) \qquad (7.5a)$$

where k' is the adsorption rate constant.

The Langmuir adsorption isotherm is given by

$$q_s = \frac{bq_m c_s}{1 + bc_s} \qquad (7.6a)$$

where b (m^3/kg) and q_m (kg/kg) are Langmuir isotherm constants.

The shape of the breakthrough curve is given by equations (7.13) and (7.14)

$$\frac{k'c_o}{q_0}(t_2 - t_1) = \frac{1}{\lambda}\ln\left(\frac{\phi_2(1-\phi_1)}{\phi_1(1-\phi_2)}\right) + \ln\left(\frac{1-\phi_2}{1-\phi_1}\right) \quad (7.13)$$

$$k' = \frac{3k_f}{R\rho(1-\varepsilon)}, \quad \phi_1 = \frac{c_1}{c_o}, \quad \phi_2 = \frac{c_2}{c_o}, \quad \lambda = 1-\beta, \quad \beta = 1 - \frac{q_0}{q_m} \quad (7.14)$$

where q_0 = solid phase concentration in equilibrium with c_0.

The velocity of movement of the centre of the adsorption or concentration front is given by

$$\left[\frac{\partial z}{\partial t}\right]_c = \frac{v_i}{1 + \frac{(1-\varepsilon)}{\varepsilon}\rho\frac{\partial q}{\partial c}} \quad (7.15)$$

The external mass transfer coefficient k_f is related to the Sherwood number Sh via the equation

$$k_f = \frac{Sh.D_{AB}}{2R} \quad (7.16)$$

where D_{AB} is the diffusion coefficient of iron(II) in water (m^2/s).

The Sherwood number Sh was computed using the mass transfer correlation given by Wilson and Geankoplis (1966) for fixed beds of spherical particles with bed porosity between 0.35 and 0.75 given by

$$\mathrm{Sh} = \frac{1.09}{\varepsilon}\mathrm{Re}^{1/3}\mathrm{Sc}^{1/3} \quad \text{for} \quad 0.0016 < \mathrm{Re} < 55 \quad \text{and} \quad 165 < \mathrm{Sc} < 70600 \quad (7.17)$$

where Re and Sc are the Reynolds number and Schmidt number respectively given by the following équations

$$\mathrm{Re} = \frac{2Rv_0}{\nu} \qquad \mathrm{Sc} = \frac{\nu}{D_{AB}} \quad (7.18)$$

where v_o = average superficial velocity (m/s) and < = kinematic viscosity (m^2/s).

2. Linear Driving Force Model

The Linear Driving Force (LDF) model, also known as a 'lumped parameter model' is a simplified approach which uses an overall kinetic rate constant (covering both internal and external mass transfer coefficients) for the adsorption rate equation. It is postulated that the uptake rate of adsorbate by an adsorbent is linearly proportional to the driving force, defined as the difference between the surface concentration and the average adsorbed phase concentration (Peel and Benedek 1981; Tien 1994). Mathematically, the solid phase mass balance is given by the linear driving force equation:

$$\frac{\partial q}{\partial t} = k(q_s - q) \tag{7.5b}$$

where k is LDF kinetic rate constant (s^{-1}).

Using mass balance and the Freundlich isotherm equation, the LDF equation for a batch system can be transformed in terms of concentration as

$$c = (c_0 - c_s)e^{-kt} + c_s \tag{7.19}$$

In logarithmic form,

$$ln\left(\frac{c - c_s}{c_0 - c_s}\right) = -kt \tag{7.20}$$

The LDF rate constant k can, therefore, be obtained from the kinetic data of batch adsorption experiments (Heijman *et al.* 1999).

In this case, the axial dispersion in the filtration columns was also included and the Freundlich isotherm equation was used for equilibrium. The Freundlich isotherm equation is given by

$$q_s = K\,c_s^{\,n} \tag{7.6b}$$

where K and n are constants for the Freundlich isotherm. These constants are determined from the adsorption equilibrium experiments.

The model equations (7.6) to (7.14) were solved numerically using the PDESOL software to obtain the iron(II) breakthrough curves.

3. Plug Flow Homogeneous Surface Diffusion Model

The Plug Flow Homogeneous Surface Diffusion Model (PFHSDM), developed for modelling adsorption in activated carbon, uses Freundlich isotherm parameters K and n and mass transfer coefficients k_f and D_s for the prediction of the breakthrough (Liu and Weber 1981; Hand *et al.* 1983; Sontheimer *et al.* 1988). The PFHSDM considers the dual mass transfer mechanism of film diffusion across the hydrodynamic boundary layer of the solid and intraparticle resistance in the form of surface diffusion. It presumes that surface diffusion is the prominent intraparticle mass transfer mechanism and is not a function of concentration. The mass transfer coefficients are determined by fitting the PFHSDM to the breakthrough of the target substance in short column tests (Weber and Liu 1980).

In this case, the axial dispersion in the filtration columns were not considered and the model equations (7.4) to (7.12) were solved numerically using the computer program written by Yuasa

(1982). This program uses a moving grid finite difference method to solve the PFHSDM equations numerically.

7.3 MATERIALS AND METHODS

7.3.1 Filter media

New (virgin) sand and iron oxide coated sand from the groundwater treatment plant Gilze, Water Supply Company North West Brabant (WNWB) in the Netherlands were used in this study. The characteristics of the filter media used are presented in Table 7.1. The specific surface areas and pore volumes were measured using BET nitrogen gas adsorption method (relative pressure range 0.05-0.25) with a Quantachrome Autosorb-6B gas adsorption analyser at 77 K at the Analytical Laboratory of the Department of Chemical Technology, Delft University of Technology. The new sand was washed in demineralised water, soaked in concentrated HCl (pH < 2) for 24 hours and again rinsed with demineralised water and dried at 40^{0}C. The iron oxide coated sand was washed gently with demineralised water to remove any deposits and dried at room temperature before using for batch and column experiments.

Table 7.1 Characteristics of the new and iron oxide coated sand used

Media	New sand	Coated sand
Grain size (mm)	0.8-1.25	0.7-1.25
Mean diameter (mm)	0.82	0.81
Average geometric surface area (m^2/g)	2.8 x 10^{-3}	2.8 x 10^{-3}
Specific surface area (BET) (m^2/g)	1.0	11.4
Surface extractable iron content (mg/g sand)	0	29-33
Pore volume (cm^3/g)		
Micropore volume (diameter < 2 nm)	0.06 x 10^{-3}	4.9 x 10^{-3}
Mesopore volume (2 nm < diameter < 50 nm)	1.24 x 10^{-3}	6.7 x 10^{-3}
Total pore volume (cm^3/g)	1.30 x 10^{-3}	11.6 x 10^{-3}

7.3.2 Batch experiments

Batch experiments were conducted under different conditions to obtain isotherm parameters and LDF kinetic rate constants. Iron(II) adsorption onto new and iron oxide coated sand was measured at different pH by conducting batch adsorption experiments at room temperature (18-20°C). The experimental set-up and procedure to determine iron(II) adsorption onto the filter media was similar to that used in an earlier study (see Chapter 2). Nitrogen gas was used for deoxygenation and the pH was adjusted using HCO_3^--CO_2 buffer. Different amounts of iron(II) sulphate ($FeSO_4.7H_2O$) stock solution (400 mg/l, pH <2) were introduced into the closed reactor containing deoxygenated water and a known weight of filter media. Samples were taken periodically from the reactor until equilibrium iron concentration was reached and the iron concentrations were measured (Standard Methods 1995). Iron(II) adsorbed onto the media as a function of time was calculated from the iron mass balance. Adsorption isotherm constants were obtained by fitting isotherm models (equations 7.6a and 7.6b) to iron(II) adsorbed onto filter media at various equilibrium iron concentrations.

The LDF rate constants (k) for both filter media at different pH were calculated by fitting the equation (7.20) to kinetic (time versus iron concentration) data of the batch adsorption experiments. The constant k is given by the slope of the straight line when $\ln(c-c_s)/(c_o-c_s)$ is plotted against time.

7.3.3 Column experiments

Laboratory-scale column experiments were conducted using a 25 mm ID, 55 cm high plexiglass column to obtain iron(II) breakthrough profiles. The schematic diagram of the experimental set-up is shown in Fig. 7.1. The depth of the filter media was 50 cm. Nitrogen gas was used to prepare the anoxic feed water. Model groundwater used in the experiments was prepared by mixing $FeSO_4.7H_2O$ stock solution (90 mg/l, pH <2, maximum age 3 days) in deoxygenated demineralised water buffered with $NaHCO_3$ to achieve an initial iron(II) concentration of 4 mg/l and an alkalinity of 100 mg/l of HCO_3^-. The pH was adjusted with HCl and NaOH. The column experiments were conducted at three different pH values i.e. 6.0, 6.5 and 7.0. The filtration rate was 8 m/h for new sand and 4 m/h for coated sand. The pH and oxygen concentration of the feed water were monitored continuously with pH meter (WTW pH 323) and oximeter (Orbisphere 3650 - detection limit 5 :g/l) to ensure constant pH and anoxic conditions.

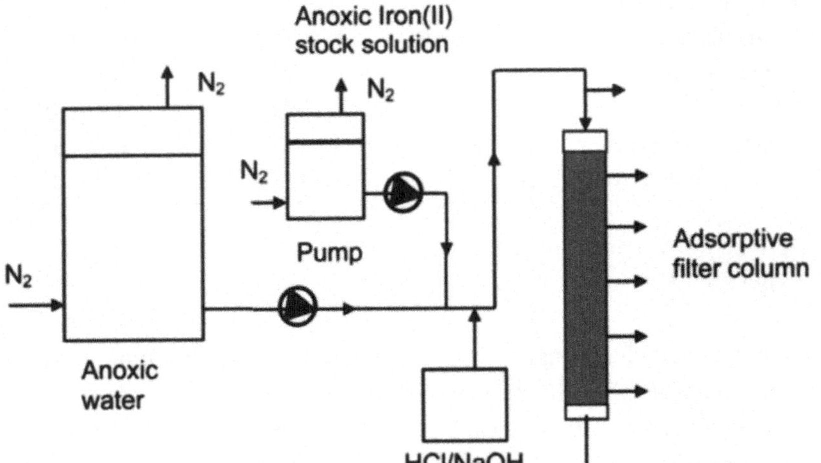

Fig. 7.1 Schematic diagram of the column experimental set-up

Before starting the filter run, the columns were brought to equilibration by recirculating the demineralised water adjusted to the pH and alkalinity of the subsequent run through the columns for about 15 minutes. The feed water and filtrate iron concentrations were measured periodically during the filter run to obtain the relative iron concentration (c/c_0) breakthrough profiles. In the case of coated sand, column experiments were terminated after a run time of 10 hours.

Determination of dispersion coefficients: The conductivity (NaCl) breakthrough experiments (without iron(II)) were conducted separately for the 50 cm-long columns equipped with new and iron oxide coated sand to assess the extent of axial dispersion in the packed bed. Concentrated sodium chloride solution (1 g NaCl/l) was introduced into the column containing filter media and the conductivities of the samples taken periodically were measured until complete breakthrough. Axial dispersion coefficients D_L were then determined by solving the model equations (7.4) to (7.12) numerically using the PDESOL software, after equating the adsorption term ($\partial q/\partial t$) to zero. The dispersion coefficients obtained in this way were used to predict the iron(II) breakthrough in the adsorptive filter columns with the LDFM.

Determination of mass transfer coefficients: Short column experiments were conducted separately for both media at pH 6.5 using the same model groundwater to obtain the external mass transfer coefficient (k_f) and surface diffusion coefficient (D_s). The bed depths of the short columns were selected after conducting several trial experiments in order to get the complete breakthrough curve within a short time. This is a requirement in these short column tests. The operational parameters for short column tests are summarized in Table 7.2.

Table 7.2 Process conditions for short column experiments

Media	Bed depth (cm)	Superficial velocity (m/h)
New sand	7	8
Coated sand	4	4

The mass transfer coefficients k_f and D_s were obtained by fitting the PFHSDM to the iron breakthrough data of short column experiments using the computer program written by Yuasa (1982). These coefficients were later used for the prediction of breakthrough in long columns.

7.4 RESULTS AND DISCUSSION

7.4.1 Isotherm parameters

Adsorption isotherm constants obtained by fitting Freundlich and Langmuir isotherm models to batch equilibrium experimental results are presented in Table 7.3. Batch results showed that adsorption of iron(II) onto both new and iron oxide coated sand gave "favourable" adsorption isotherms ($n < 1$). Systems with favourable adsorption isotherms show "constant pattern" behaviour with respect to the form of mass transfer front or removal profiles along the depth of the filter, i.e. the breakthrough curve retains a constant shape (Ruthven 1984; Wankat 1990; Tien 1994). Therefore, the Constant Pattern Model was the first to be selected for the prediction of iron breakthrough in adsorptive filter columns.

Table 7.3 Adsorption isotherm parameters

Media	Isotherm	Parameter	pH 6.0	pH 6.5	pH 7.0
New sand					
	Freundlich	K	6.4×10^{-6}	6.9×10^{-6}	1.1×10^{-5}
		n	0.54	0.73	0.70
		r^2	0.95	0.98	0.98
	Langmuir	q_m	3.1×10^{-5}	6.2×10^{-5}	6.3×10^{-5}
		b	0.22	0.115	0.208
		r^2	0.97	0.98	0.96
Coated sand					
	Freundlich	K	1.2×10^{-4}	1.7×10^{-4}	3.3×10^{-4}
		n	0.49	0.54	0.43
		r^2	0.96	0.99	0.98
	Langmuir	q_m	4.1×10^{-4}	7.5×10^{-4}	9.2×10^{-4}
		b	0.374	0.26	0.514
		r^2	0.94	0.98	0.93

Units: K [(g/g)/(g/m^3)^n], n [-], q_m [g/g], and b [m^3/g].

7.4.2 Model predictions

1. Constant Pattern Model

In Constant Pattern Model (CPM), internal mass transfer and axial dispersion in the column were neglected. The film mass transfer coefficients (k_f) for the new and iron oxide coated sand were computed from equations (7.16) to (7.18). Parameters used and the results of this calculation are summarised in Table 7.4.

Kinematic viscosity of water υ = 10^{-6} m^2/s (at 20°C)
Diffusion coefficient D_{AB} for iron(II) = 0.72×10^{-9} m^2/s (CRC Handbook 1996)

Table 7.4 Film mass transfer coefficients and parameters used for calculation

	New sand	Coated sand
Particle diameter 2R (m)	8.2 x 10^{-4}	8.1 x 10^{-4}
Superficial velocity vo (m/h)	8	4
Fixed bed porosity ,	0.4	0.4
Reynolds number Re	1.82	0.90
Schmidt number Sc	1391	1391
Sherwood number Sh	37.2	29.4
Film mass transfer coefficient k_f (m/s)	3.26 x 10^{-5}	2.61 x 10^{-5}

Figures 7.2 and 7.3 compare the iron breakthrough curves predicted from equations (7.13) to (7.18) and the experimental data at different pH values for the filter columns with new and iron oxide coated sand respectively. It was observed that i) the iron breakthrough in the experimental columns occurred much earlier than predicted by the model, and ii) at the later stage, the increase of filtrate iron concentrations in experimental columns was considerably slower while the model predicted a sharp increase in iron concentration. This indicates that the overall mass transfers in adsorptive iron removal filters are likely to be much slower than that predicted by the CPM. In particular, the internal mass transfer is most likely a limiting factor, as this is not taken into account in the CPM.

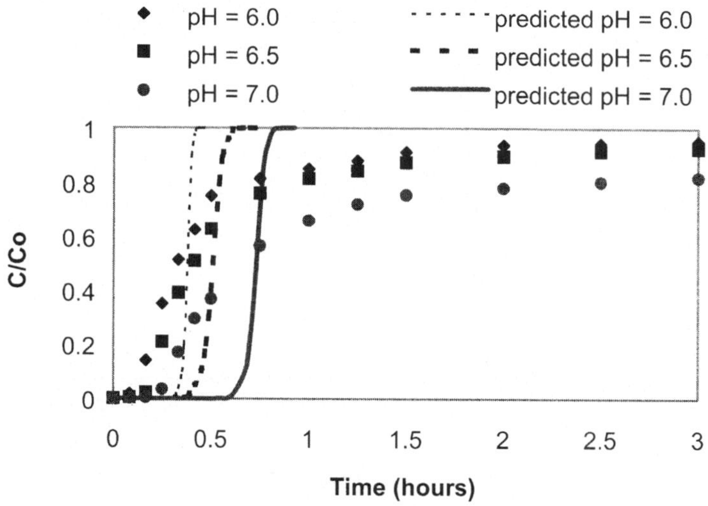

Fig. 7.2 Predicted and experimental iron breakthrough curves for the filter column with new sand at different pH (filtration rate = 8 m/h, initial iron concentration = 4 mg/l)

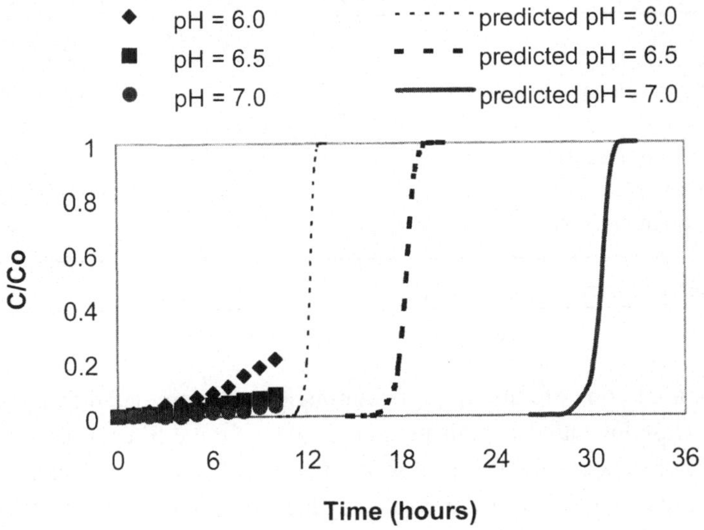

Fig. 7.3 Predicted and experimental iron breakthrough curves for the filter column with iron oxide coated sand at different pH (filtration rate = 4 m/h, initial iron concentration = 4 mg/l)

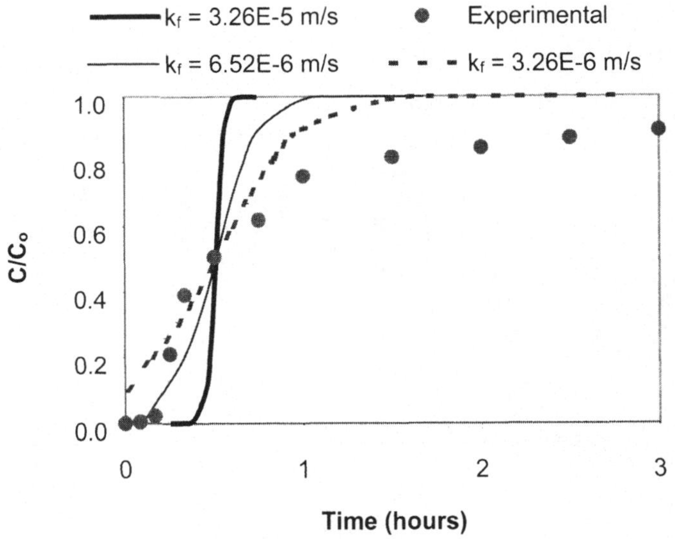

Fig. 7.4 Experimental and predicted iron breakthrough curves for the filter column with new sand for different values of external mass transfer coefficient k_f (filtration rate = 8 m/h, initial iron concentration = 4 mg/l, pH = 6.5)

Figure 7.4 shows the experimental and predicted iron breakthrough curves for the filter column with new sand for different values of external mass transfer coefficient k_f. It is clear from Fig. 7.4 that even when k_f is decreased by 5 or 10 times, there are still considerable differences between model predictions and experimental data. This implies that besides external mass transfer, there are also some other factors that affect the adsorptive iron removal process. Nitrogen physisorption measurements using the BET theory demonstrated that both new and coated sand have considerable internal porosity and high internal surface area (Table 7.1). The iron(II) breakthrough might be influenced by adsorption of iron(II) onto the internal surfaces, or the internal mass transfer might play a role in the removal process. Therefore, the consideration of external film mass transfer alone was not sufficient to model accurately the process of adsorptive iron removal.

2. Linear Driving Force Model

As the CPM was not satisfactory, the LDFM including the axial dispersion in the columns was applied in order to get better predictions of iron breakthrough in adsorptive filter columns. As the ratio of column diameter to grain diameter was 31, the dispersion due to wall effect was probably negligible (Leclerc 1975). Notwithstanding this, it is useful to quantify the effect of axial dispersion due to other phenomena on the iron(II) breakthrough in adsorptive filters.

The LDF overall kinetic rate constant (k) was obtained by fitting equation (7.20) to kinetic data (iron concentration versus time) of batch adsorption experiments. A typical graph for the determination of the rate constant k is presented in Fig. 7.5, and the LDF kinetic rate constants for new and coated sand are summarised in Table 7.5.

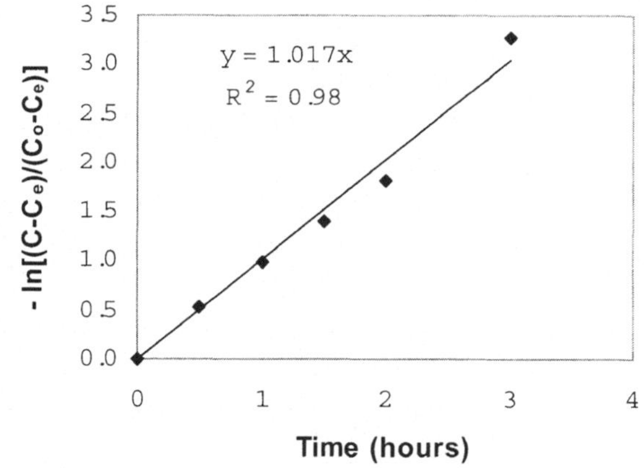

Fig. 7.5 Determination of the LDF kinetic rate constant k from the time-iron concentration data of batch experiments (pH = 6.5, new sand, initial iron concentration = 4 mg/l). Constant k is given by the slope of the straight line.

Table 7.5 LDF rate constants at different pH

Media	LDF rate constant k (/s)		
pH	6.0	6.5	7.0
New sand	2.6×10^{-4}	2.8×10^{-4}	2.9×10^{-4}
Coated sand	1.7×10^{-4}	1.8×10^{-4}	2.0×10^{-4}

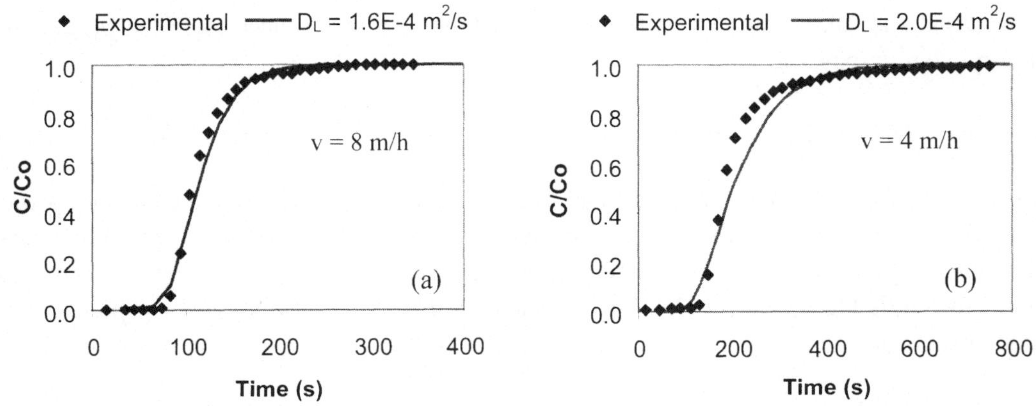

Fig. 7.6 Predicted and experimental conductivity breakthrough curve for the filter columns with (a) new sand and (b) iron oxide coated sand.

Axial dispersions in the experimental columns were measured by performing the conductivity (NaCl) breakthrough experiments. Figure 7.6 shows the typical experimental and predicted (best fitting) conductivity breakthrough curves for the 50 cm filter column with new and coated sand. Using the model equations (7.4) to (7.12) for the system without adsorption, the dispersion coefficients for new sand and iron oxide coated sand, under given process conditions, were determined to be $1.6 \pm 0.1 \times 10^{-4}$ m^2/s and $2.0 \pm 0.1 \times 10^{-4}$ m^2/s, respectively.

Figures 7.7 and 7.8 compare the LDFM predictions (including axial dispersion) with the experimental iron breakthrough data. Figure 7.7 shows that the LDFM fairly accurately predicted the iron(II) breakthrough in the adsorptive column with new sand. The LDFM predictions for all three pH values were very close to the experimental results. In the case of coated sand (Fig. 7.8), the LDFM predictions of iron breakthrough were closer to the experimental data than that of the CPM. However, there was still a considerable difference between the experimental data and the LDFM predictions in the case of coated sand.

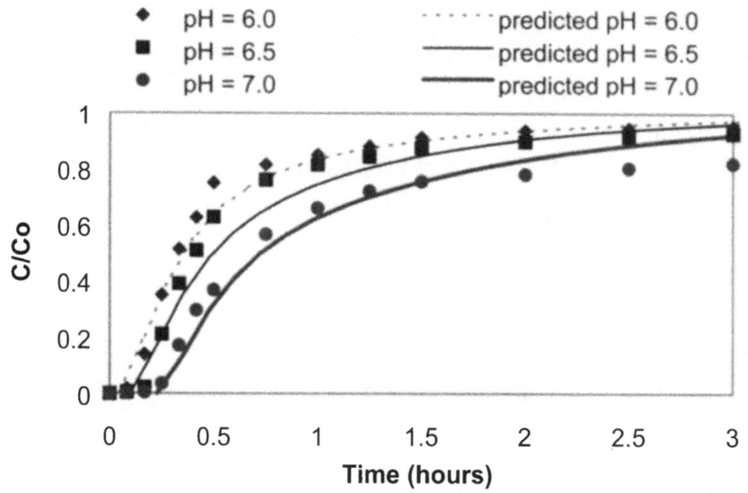

Fig. 7.7 Predicted (LDFM) and experimental iron breakthrough curves for the filter column with new sand (filtration rate = 8 m/h, initial iron concentration = 4 mg/l)

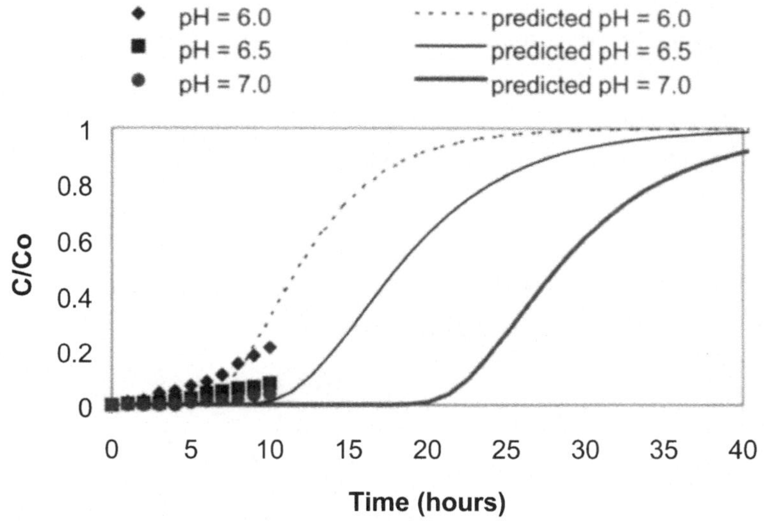

Fig. 7.8 Predicted (LDFM) and experimental iron breakthrough curves for the filter column with iron oxide coated sand (filtration rate = 4 m/h, initial iron concentration = 4 mg/l)

The dispersion in a fixed bed adsorber depends on the column geometry and the flow conditions. The deviation from the ideal plug flow condition is caused by short-circuiting, by recycling, or by the presence of stagnant zones. The dispersion number (D_L/v_iL) is the measure of the degree of variation from the ideal plug flow condition (0 for plug flow and infinity for completely mixed). Depending upon the dispersion number, the extent of dispersion in the columns is characterised as small ($D_L/v_iL = 0.002$), intermediate ($D_L/v_iL = 0.025$), and large ($D_L/v_iL = 0.2$) (Levenspiel 1972; Weber 1972). Under the applied conditions, the dispersion numbers were calculated as 0.06 and 0.14 for the experiments with new and coated sand respectively. The effect of dispersion is likely to be more pronounced at the lower velocity and for the short columns. At higher velocity and for longer columns, the effect of dispersion is likely to be negligible (Weber 1972; Bhattarai and Griffin 1999). It is consequently likely that the some dispersion effect was observed because a relatively short (50 cm) experimental column was used in this study. In practice, filter bed depth and filtration rate range between 1.5-2.0 m and 5-15 m/h respectively. Therefore, the dispersion effect is likely to be rather small in full-scale filters.

3. Plug Flow Homogeneous Surface Diffusion Model

Since both models, the CPM and the LDFM, do not take into account internal and external mass transfer, the PFHSDM was applied to predict iron breakthroughs in filter columns with new and iron oxide coated sand. The PFHSDM separately considers both external mass transfer and internal mass transfer. The mass transfer coefficients (k_f and D_s) obtained by fitting the PFHSDM to short column iron breakthrough data are summarised in Table 7.6. The experimental data and predicted short column breakthrough curves for new and iron oxide coated sand are presented in Figs. 7.9a and 7.9b respectively.

Table 7.6 Mass transfer coefficients from short column tests

Media	k_f (m/s)	D_s (m^2/s)
New sand	$2.27 \pm 0.20 \times 10^{-5}$	$2.30 \pm 0.20 \times 10^{-11}$
Coated sand	$1.65 \pm 0.13 \times 10^{-5}$	$0.13 \pm 0.01 \times 10^{-11}$

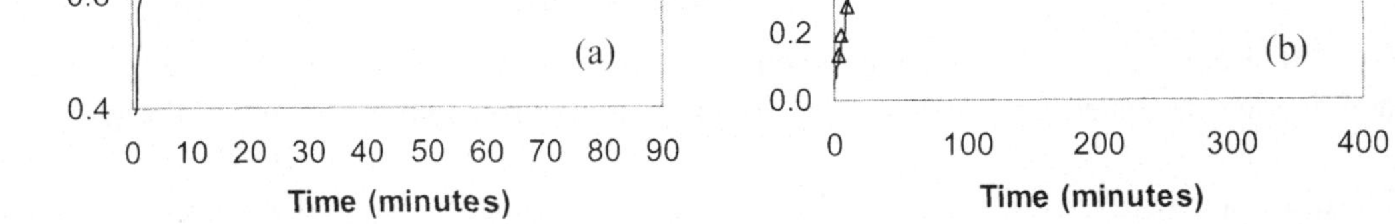

Fig. 7.9 Short column iron breakthrough curves for (a) new sand and (b) iron oxide coated sand (initial iron concentration = 4 mg/l, pH = 6.5)

The external mass transfer coefficients (k_f) for new and coated sand obtained from the short column experiments (Table 7.6) were 70% and 63% respectively of the external mass transfer coefficients calculated theoretically from mass transfer correlations (Table 7.4). This is probably due to the actual iron(II) adsorption capacity in the filter columns being smaller than the adsorption capacity obtained from the batch adsorption experiments.

Figures 7.10 and 7.11 present the iron breakthrough curves for the 50 cm column predicted from the PFHSDM (neglecting axial dispersion) together with the experimental data at different pH values for new and coated sand respectively. Figure 7.10 shows that the PFHSDM fairly accurately predicted the initial increase in filtrate iron concentration in the adsorptive column with new sand. However, with the further progress of the filter run, there were considerable differences between the model predictions and experimental data. At all three pH values studied, the filtrate iron concentration did not reach the feed water iron concentration but the breakthrough curve was "broadening". This may be due to some continuous removal of iron in the column, which was achieved through either ion exchange or iron(II) oxidation by traces of oxygen possibly present in the pores of the filter media and in the feed water.

Fig. 7.10 Predicted (PFHSDM) and experimental iron breakthrough curves for the filter

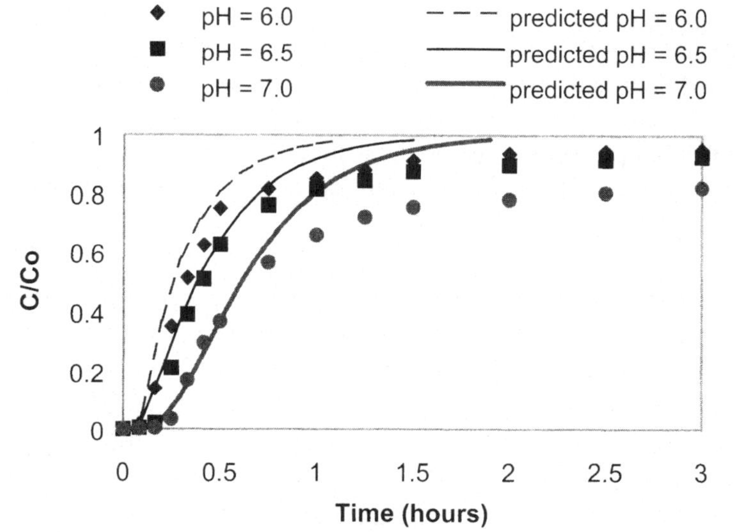

column with new sand (filtration rate = 8 m/h, initial iron concentration = 4 mg/l)

Figure 7.11 shows that the PFHSDM provides a somewhat better prediction of the iron breakthrough in adsorptive filters with iron oxide coated sand than the CPM. However, the initial increase of iron concentration in the experimental column occurred much earlier than the model predicted. Further increase in filtrate iron concentration observed in the experiments, however, was very slow compared to a sharp increase predicted by the model. There is not sufficient experimental data to compare the agreement of the model predictions with the later part of the breakthrough curve. For columns with iron oxide coated sand, the LDFM and the PFHSDM predictions of iron breakthrough were very similar.

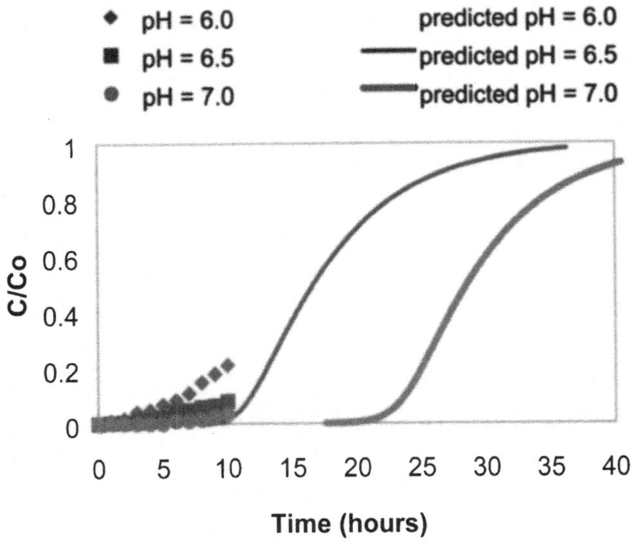

Fig. 7.11 Predicted (PFHSDM) and experimental iron breakthrough curves for the filter column with coated sand (filtration rate = 4 m/h, initial iron concentration = 4 mg/l)

The axial dispersion was not considered initially in the PFHSDM because the model considers "plug flow" conditions and some effect of axial dispersion was already included in the mass transfer coefficients k_f and D_s determined from the short column experiments. Taking into account the effect of dispersion, it was found that for the measured values of k_f and D_s for new and iron oxide coated sand, there were no significant changes in the PFHDSM model predictions with or without inclusion of the axial dispersion.

In agreement with the results of batch adsorption experiments and filter column runs, for both the media, all three models used predicted the enhanced adsorptive iron removal at higher pH. The "broadening" of the breakthrough curves in the case of new sand might be due to the presence of some traces of oxygen in the feed water (5 :g/l - the detection level of the oxygen meter used) which might have continuously regenerated some of the iron(II) adsorption sites, thus resulting in an extra capacity (5 :g/l of oxygen can oxidise about 0.04 mg/l of iron(II)). In the case of coated sand, however, the contribution of this effect to overall removal is likely to be minimal because its iron(II) adsorption capacity is much higher than that of new sand.

It should be noted that, in the case of the short column tests, adsorption on the outer or external pore spaces probably plays a major role and determines the mass transfer coefficients. In the case of new sand, external pores are relatively larger and more in number, so the effect of internal pores is most likely minimal. However, in the case of coated sand the internal pores are likely to be considerably more in number and relatively smaller than those in new sand. Therefore, it is likely that the actual surface diffusion coefficient D_s in the case of coated sand is much smaller than that measured from the short column tests. Hence, the actual iron breakthrough in adsorptive filters with iron oxide coated sand might be much slower than that predicted by the model.

Furthermore, the PFHSDM assumes that the adsorbent is homogeneous and the surface diffusion is the dominant mass transfer mechanism. However, the iron oxide coated sand is not homogeneous but consists of a sand core and an outer layer of iron oxide coating. Additionally, the new sand grain as well as iron oxide coating are not homogeneous but consist of pores of different sizes, and the iron oxide coating is not uniform throughout the sand grain (see Chapter 4). This is likely one of the reasons for the difference in the PFHSDM predictions and experimental results.

7.4.3 Sensitivity analysis of model parameters

The sensitivity of the LDFM and the PFHSDM predictions with the change in the values of model parameters, (specifically adsorption capacity, mass transfer coefficients, and dispersion coefficients) were analysed. This was done to check whether the differences between the experimental results and model predictions were due to possible errors associated with parameter determinations. While analysing the sensitivity of one parameter, the other parameters were kept constant and their respective measured values were used.

a) LDFM

Figures 7.12 and 7.13 present the sensitivity of the LDFM breakthrough predictions for the filter column with new sand and iron oxide coated sand for different values of the LDF rate constant k and the dispersion coefficient D_L. For both the media, iron breakthroughs were earlier when k was decreased or D_L was increased. The LDFM predictions for filter columns with coated sand, within the range of k and D_L tested, did not fit the experimental breakthrough values. This indicates that apart from mass transfer, some other mechanism may also be influencing the iron(II) adsorption capacity and, consequently, the shape of the iron breakthrough curve.

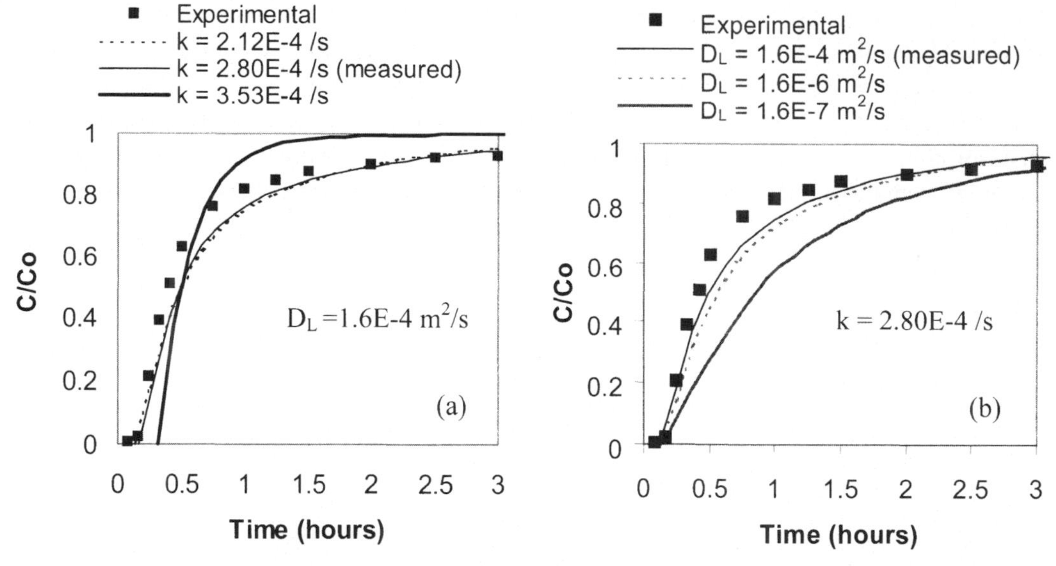

Fig. 7.12 Effect of (a) k and (b) D_L on the LDFM predictions of iron breakthrough for the filter column with new sand (pH = 6.5, initial iron concentration = 4 mg/l, filtration rate = 8 m/h)

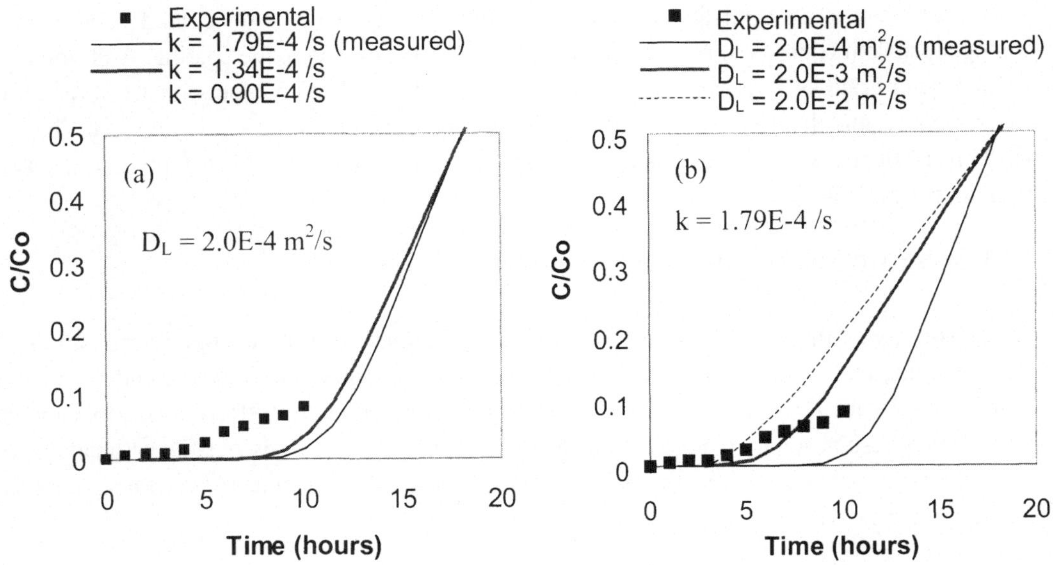

Fig. 7.13 Effect of (a) k and (b) D_L on the LDFM predictions of iron breakthrough for the filter column with coated sand (pH = 6.5, initial iron concentration = 4 mg/l, filtration rate = 4 m/h)

b) PFHSDM

Figures 7.14 and 7.15 present the effect of mass transfer coefficients (k_f, D_s) and adsorption capacity (K) on iron breakthrough curves for the filter columns with new sand and coated sand. Figure 7.14a shows that there was no significant change in the shape of the breakthrough curve with ± 50% change in k_f, whereas some alteration of D_s and adsorption capacity K influenced both the shape of the curve and breakthrough time significantly. In the case of iron oxide coated sand, a decrease in k_f and D_s influenced the initial part of the curve and gave early iron breakthrough (Fig. 7.15). As expected, the breakthrough time was more sensitive to the changes in adsorption capacity K. However, in the case of coated sand, the breakthrough curve was very different from the model predictions with different sets of parametric value, indicating that the difference was not due to possible errors associated with the parameter determinations.

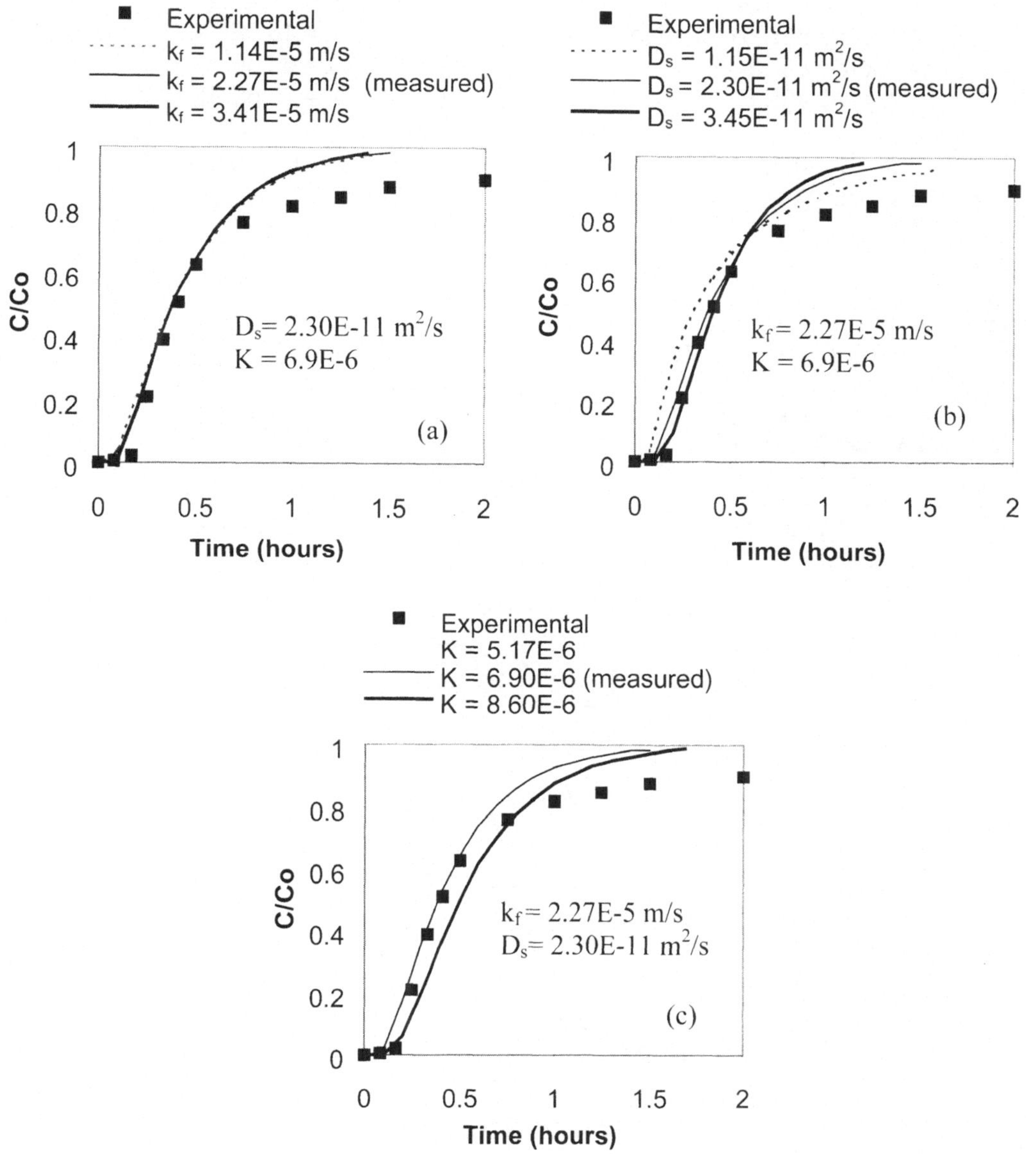

Fig. 7.14 Effect of (a) k_f, (b) D_s and (c) K on the PFHSDM predictions of iron breakthrough for filter column with new sand (pH = 6.5, initial iron concentration = 4 mg/l, filtration rate = 8 m/h)

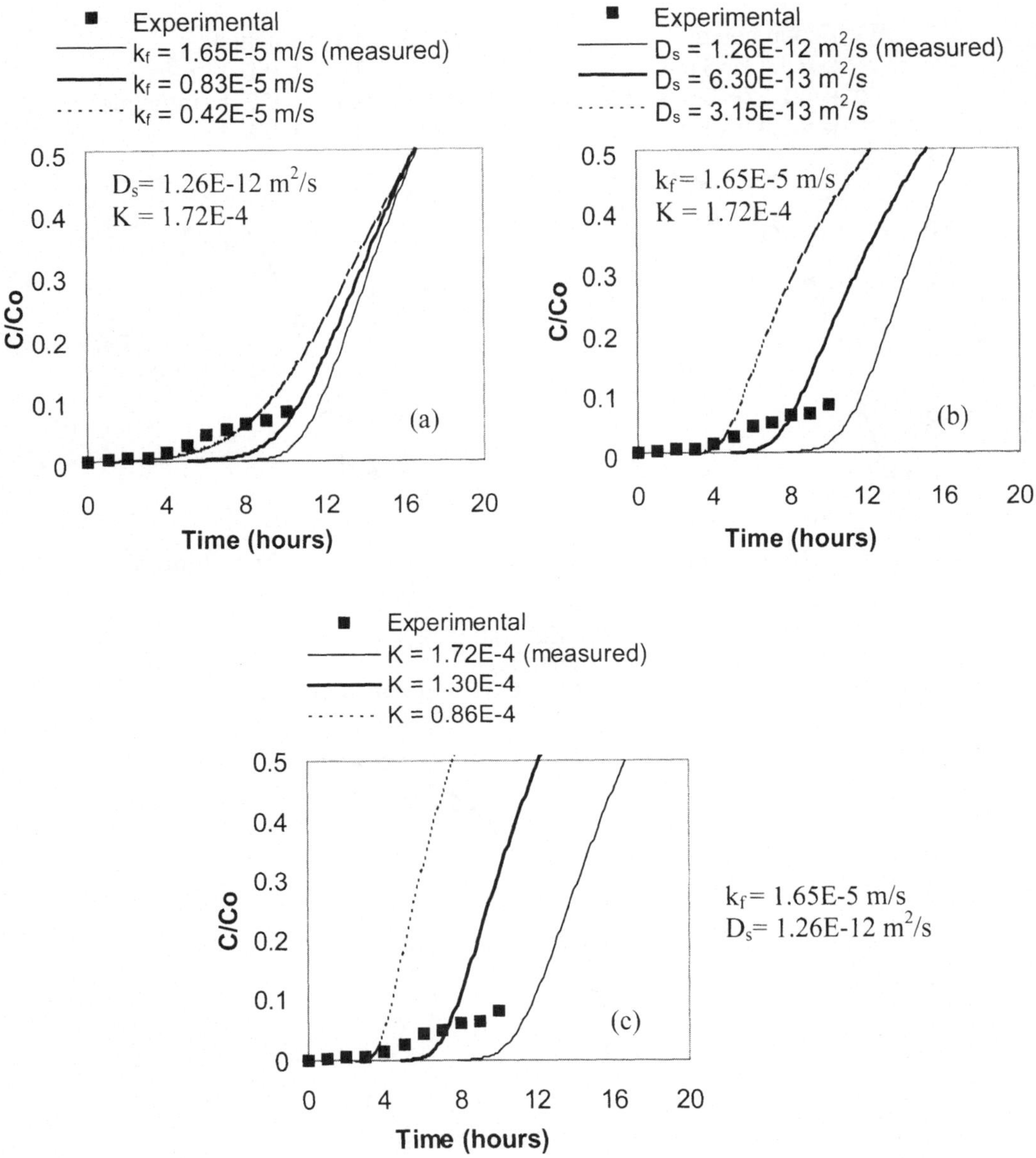

Fig. 7.15 Effect of (a) k_f, (b) D_s and (c) K on the PFHSDM predictions of iron breakthrough for the filter column with coated sand (pH = 6.5, initial iron concentration = 4 mg/l, filtration rate = 4 m/h)

Based on the experimental results, model predictions and sensitivity analysis, the following can be summarised:

- The initial iron breakthrough in the filter column with new sand can be predicted reasonably accurately with both the LDFM and the PFHSDM. The subsequent development of the curve showed a remarkable difference between the experimental results and the predictions of both models.

- The iron breakthrough curves for the filter column with iron oxide coated sand has very different shapes than that predicted by both the LDFM and the PFHSDM.
- The variation in iron(II) adsorption capacity has a more pronounced effect than internal and external mass transfers, and axial dispersion.

Taking these observations into account, it was hypothesised that the pH drop in the pores of the filter media due to adsorption of iron(II) is the reason behind the discrepancies between the measured and the predicted iron breakthrough curves in adsorptive filter columns.

7.4.4 pH drop in the pores of filter media with iron(II) adsorption

In the case of filters with iron oxide coated sand, initial increases in filtrate iron concentration were much earlier than that predicted by both the models (PFHSDM and LDFM). Further increases in iron concentrations were, however, much slower than the model predictions. The difference between the model predictions and the experimental results in this case was probably due to the *"pH drop"* in the pores of the filter media due to iron(II) adsorption. Adsorption of iron(II) onto filter media is accompanied by release of the proton (H^+) (Takai 1973; Barry *et al.* 1994), associated with a local pH drop in the pore spaces and on the surface of the filter media. This pH drop results in a temporary decrease in the iron(II) adsorption capacity. The extent of pH drop depends on the density of adsorption sites and the diameter of the pores. The calculated densities of iron(II) adsorption sites under the given conditions are presented in Appendix 7.1. Considering the filter media pore space as a closed system (no transport of H^+ and HCO_3^- out of the pore volume), the theoretical calculations of local pH drop in the pores of the filter media with iron(II) adsorption were made (Appendix 7.2). The BET Nitrogen adsorption measurements showed that the coated sand tested had high pore volume (Table 7.1) and contained both micropores (diameter < 2 nm) and mesopores (2 nm < diameter < 50 nm). About 0.005 cm^3/g of pore volume were obtained from micropores and about 0.007 cm^3/g of pore volume from mesopores. Figure 7.16 shows that this local pH drop in the pores of the filter media with iron(II) adsorption is substantial for a pore diameter up to 5 μm, which consequently reduces the iron(II) adsorption capacity significantly.

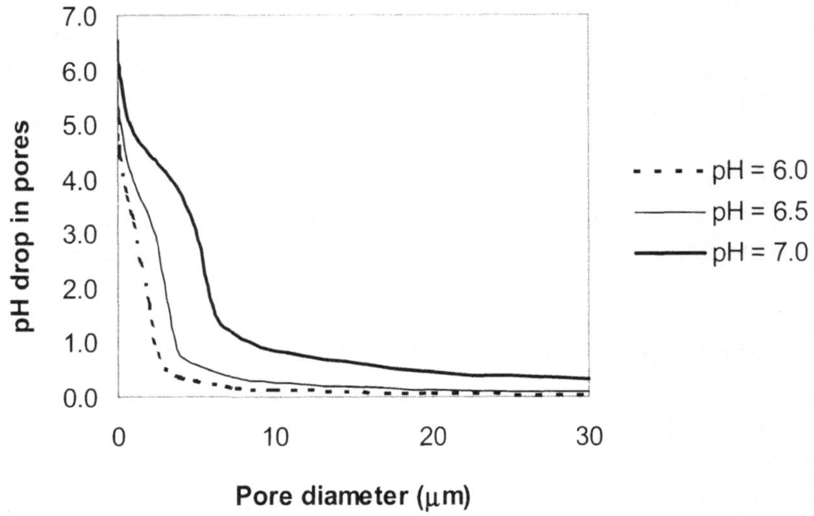

Fig. 7.16 pH drop in pores of iron oxide coated sand due to iron(II) adsorption at different pH as a function of pore diameter (HCO_3^- = 100 mg/l, iron(II) concentration = 4 mg/l).

The proton released due to iron(II) adsorption, however, will react with the alkalinity (bicarbonate ions) present in water e.g. diffusing from the bulk solution, thus compensating the pH change and regaining the iron(II) adsorption capacity. The diffusion of H^+ ions outside the pores and/or transport of HCO_3^- ions inside the pores are time and concentration dependent processes governed by flow conditions, pore geometry, and diffusion coefficients of H^+ and HCO_3^- ions. Therefore, it is likely that, initially, the iron breakthrough is faster than the model predictions as the adsorption capacity is lower (due to local pH drop). Later on, iron breakthrough is much slower as the adsorption capacity is much higher than before due to an increase of pH and also because the adsorption capacity which was under-utilised initially, contributes to the overall adsorption.

This "pH effect" on iron(II) adsorption is likely to be more pronounced at lower pH as the adsorption capacity decreases substantially with the decrease of pH. Additionally, this effect would be much less pronounced in the case of new sand as adsorption capacity is considerably smaller and, hence, the local pH drop with initial iron(II) adsorption is much lower. Furthermore, the pores in new sand are likely to be mainly close to the surface and relatively bigger than in iron oxide coated sand. Hence, the diffusion of H^+ and HCO_3^- is faster, and the pH drop would be minimal. In the case of coated sand, pore sizes are most likely much smaller and pores are much deeper inside the grain than those in new sand. Therefore, the pH effect is likely to be much more pronounced.

As mentioned earlier, in the short column tests, adsorption on the outer or external pore spaces is expected to be dominant, and this determines the mass transfer coefficients. The diffusion of iron(II) into the internal pores and subsequent pH drop due to iron(II) adsorption are also time-dependent. The internal pores are relatively much smaller than the external pores. The pH effect increases with a decrease in pore sizes. Therefore, it is likely that the "pH effect" is less pronounced in the case of short column tests than in the long columns.

It should also be noted that in reality the actual pH drop is not so pronounced because, with the adsorption of iron(II), the pH starts to drop, consequently the adsorption capacity also decreases and all the adsorption sites available at the beginning are not occupied.

7.5 CONCLUSIONS

The process of adsorptive iron removal in filters under anoxic conditions was modelled using adsorption isotherm parameters, mass balance and mass transfer equations. For this fixed bed adsorption model, the iron breakthroughs in adsorptive filters with new and iron oxide coated sand were predicted using three different models with increasing complexity, and the model predictions were compared with the experimental results. The following conclusions can be drawn:

- The experimental iron breakthrough in adsorptive filters occurred much earlier than the predictions of the Constant Pattern Model (CPM), which considers film (external) mass transfer only. In addition, the shapes of the predicted iron breakthrough curves were considerably different from the experimental breakthrough curves.
- The Linear Driving Force Model (LDFM), which considers an overall mass transfer combining both external and internal mass transfer, gave a good prediction for the initial breakthrough in filter columns with new sand. However, the subsequent development gave a significant deviation. For coated sand, the model predictions improved but were still quite different from the measured values.
- The Plug Flow Homogeneous Surface Diffusion Model (PFHSDM), which separately includes both film mass transfer and surface diffusion, gave a good prediction of initial breakthrough in filter columns with new sand. Further development of the curve, as in the case of the LDFM, gave a significant deviation from experimental results. For coated sand, the iron breakthrough predictions with PFHSDM were not better than with the LDFM.
- Incorporating dispersion in the LDF and the PFHSDM gave no substantial improvement in iron breakthrough predictions.
- The difference in model predictions and experimental results in the case of iron oxide coated sand was probably due to the effect of an initial pH drop in the pores with iron(II) adsorption, and a consequent decrease in iron(II) adsorption capacity.
- More accurate predictions of iron breakthrough will require quantification of the effect of pH change in the pores of the filter media with iron(II) adsorption. In addition, the effect of inhomogeneity of pore sizes, and traces of oxygen in the water and in the pores needs to be studied in more detail.

REFERENCES

Barry, R.C., Schnoor, J.L., Sulzberger, B., Sigg, L. and Stumm, W. (1994) Iron oxidation kinetics in an acidic alpine lake. *Water Research*, **28** (2), 323-333.

Bhattarai, R. R. and Griffin Jr., D.M. (1999) Results of tracer tests in rock-plant filters. *Journal of Environmental Engineering*-ASCE, **125** (2), 117-125.

CRC Handbook of Chemistry and Physics (1996) 77th Edition, CRC Press Inc.

Faust, S.D. and Aly, O.M. (1987) *Adsorption Processes for Water Treatment.* Butterworth Publishers, USA.

Hand, D.W., Crittenden, J.C. and Thacker, W.E. (1983) User-oriented batch reactor solutions to the homogeneous surface diffusion model. *Journal of Environmental Engineering*-ASCE, **109** (1), 82-101.

Hand, D.W., Crittenden, J.C. and Thacker, W.E. (1984) Simplified models for design of fixed bed adsorption systems. *Journal of Environmental Engineering*-ASCE, **110** (2), 440-456.

Heijman, S.G.J., van Paassen, A.M., van der Meer, W.G. J. and Hopman, R. (1999) Adsorptive removal of natural organic matter during drinking water treatment. *Water Science and Technology*, **40** (9), 183-190.

Michaels, A.S. (1952) Simplified method of interpreting kinetic data in fixed bed ion exchange. *Industrial Engineering Chemistry*, **44** (8), 1922–1930

Lecrec, D. (1975) Characteristics of unconsolidated filter media. In: *The Scientific Basis of Filtration.* Ives, K.J. (Editor), 151-166. Noordhoff International Publishing, Leiden, The Netherlands.

Levenspiel, O. (1972) *Chemical Reaction Engineering.* 2nd Edition, John Wiley & Sons, Inc.

Liu, K.T. and Weber, W.J. Jr. (1981) Characterization of mass transfer parameters for adsorber modeling and design. *Journal WPCF*, **53** (10), 1541–1550.

Peel, R.G. and Benedek, A. (1981) A simplified driving force model for activated carbon adsorption. *The Canadian Journal of Chemical Engineering*, **59** December, 688-692.

Ruthven, D.M. (1984) *Principles of Adsorption and Adsorption Process.* John Wiley & Sons, USA.

Schwertmann, U. and Cornell, R.M. (1991) *Iron Oxides in the Laboratory: Preparation and Characterisation.* VCH Publications, Germany.

Smith E.H. (1991) Modified solution of homogeneous surface diffusion model for adsorption. *Journal of Environmental Engineering*-ASCE, **117**, 320-338.

Snoeyink, V.L. and Jenkins, D. (1980) *Water Chemistry.* John Wiley & Sons, USA.

Sontheimer, H., Crittenden, J.C. and Summers, R. S. (1988) *Activated Carbon for Water Treatment.* AWWA-DVGW Forschungsstelle Engler Bunte Institut, Karlshue, Germany.

Standard Methods for the Examination of Water and Wastewater (1995) 19th edition, American Public Health Association/American Water Works Association/Water Environment Federation, Washington DC.

Takai, T. (1973) Studies on the mechanisms of catalytic deferrization (III) - Clarification of the mechanics of self catalytic deferrization. *Journal of Japan Water Works Association*, 467, August, 16-23.

Tien, C. (1994) *Adsorption Calculations and Modelling.* Butterworth-Heinemann, USA.

Wankat, P.C. (1990) *Rate Controlled Separations.* Elsevier Applied Science.

Weber, W.J., Jr. (1972) *Physicochemical processes for water quality control.* Wiley Interscience Publication; John Wiley & Sons Inc., USA.

Weber, W.J. Jr. and Liu, K.T. (1980) Determination of mass transport parameters for fixed bed adsorbers. *Chem. Eng. Commun.*, **6**, 264–270.

Wilson, E.J. and Geankoplis, C.J. (1966) Liquid mass transfer at very low Reynolds numbers in packed beds. *Ind. Eng. Chem. Fund.*, **5** (1), 9-14.

Yuasa, A. (1982) *A Kinetic Study of Activated Carbon Adsorption Processes.* Ph.D. Thesis, Hokkaido University, Sapparo, Japan.

APPENDIX 7.1 Surface coverage with iron(II) adsorption and adsorption site densities

Using the Freundlich isotherm equation, iron(II) adsorbed onto filter media was calculated at given pH and equilibrium iron(II) concentration. The total area covered by adsorbed iron(II) was then compared with the BET specific surface area of the filter media to obtain filter media surface coverage and iron(II) adsorption site densities.

Ionic diameter of iron(II) ion = 154 pm = 1.54×10^{-10} m (Schwertmann and Cornell 1991)
Projected area of one iron(II) ion = 2.37×10^{-20} m^2

Table 7.A.1 Calculation of surface coverage and iron(II) adsorption site densities

	pH = 6.0		pH = 6.5		pH = 7.0	
	New sand	CS Gilze	New sand	CS Gilze	New sand	CS Gilze
Equilibrium iron(II) concentration Ce (mg/l)	4	4	4	4	4	4
Iron(II) adsorbed (g/g media)	1.77E-05	2.54E-04	1.90E-05	3.64E-04	2.94E-05	6.98E-04
Moles of iron(II) adsorbed (mol/g media)	3.17E-07	4.54E-06	3.39E-07	6.51E-06	5.27E-07	1.25E-05
No of iron(II) ions in 1 mole	6.02E+23	6.02E+23	6.02E+23	6.02E+23	6.02E+23	6.02E+23
No of iron(II) ions adsorbed	1.91E+17	2.73E+18	2.04E+17	3.92E+18	3.17E+17	7.52E+18
Area covered by iron(II) ions (m^2)	4.53E-03	6.48E-02	4.84E-03	9.29E-02	7.52E-03	1.78E-01
Total surface area -BET (m^2/g media)	1.00E+00	1.14E+01	1.00E+00	1.14E+01	1.00E+00	1.14E+01
Surface coverage with iron(II)	**0.45%**	**0.57%**	**0.48%**	**0.81%**	**0.75%**	**1.56%**
Adsorption site density (mol/m^2)	3.17E-07	3.98E-07	3.39E-07	5.71E-07	5.27E-07	1.10E-06
Adsorption site density (ion/m^2)	**1.91E+17**	**2.40E+17**	**2.04E+17**	**3.44E+17**	**3.17E+17**	**6.60E+17**

APPENDIX 7.2 Calculations of the pH changes in the pores with iron(II) adsorption

Pore diameter (m) = d_p
Pore length (m) = L_p
Internal surface area of the cylindrical pores A_p (m^2) = $\pi d_p L_p$
Pore volume V_p (m^3) = $0.25\ \pi\ d^2\ L$

Density of iron(II) adsorption sites (mol iron(II)/m^2) = σ
Total iron(II) adsorbed (mol) = $\sigma . A_p = \sigma . \pi d_p L_p$

As two moles of H^+ are released per mole of iron(II) adsorbed,
Total H^+ released (mol) = $2\ \sigma . \pi d_p L_p$

Concentration of H^+ released in pores due to iron(II) adsorption ($[H^+_{rel}]$) $= 2\ \sigma . \pi d_p L_p / V_p$
$= 8\ \sigma / d_p$ mol/m^3
$= 8 \times 10^{-3}\ \sigma / d_p$ mol/l

When there is no transport of H^+ ions outside and HCO_3^- ions inside the given pore volume, the pH drop in the pores is determined by the initial pH and alkalinity (HCO_3^-) in the pores.

Depending on the density of adsorption sites and pore diameter, there are two possible cases:

I) when $[H^+_{rel}]$ > alkalinity, all alkalinity will be consumed by H^+ and the final pH is determined by the remaining H^+ ions in the pores.

II) when $[H^+_{rel}]$ < alkalinity, H^+ released will convert HCO_3^- to CO_2 and the final pH is determined by the HCO_3^- - CO_2 equilibria.

$$K_1 = [H^+]\,[HCO_3^-] / [CO_2] \tag{7.A.1}$$

$pK_1 = 6.38$ at 20°C (Snoeyink and Jenkins 1980)

Sample calculations

pH drop in pores when iron(II) is adsorbed onto iron oxide coated sand at pH = 7.0, and equilibrium iron(II) concentration = 4 mg/l,

Initial pH = 7.0
Initial $[H^+] = 1.0 \times 10^{-7}$ mol/l
Initial $[HCO_3^-]$ = 100 mg/l = 1.64×10^{-3} mol/l

Density of iron(II) adsorption sites σ (mol iron(II)/m^2) = 1.10×10^{-6} (See Appendix 7.1)

Case (I)
Pore diameter = 25 nm = 25 x 10^{-9} m
Concentration of H^+ released = 0.352 mol/l

As $[H^+_{rel}] > [HCO_3^-]$, all HCO_3^- will be converted to CO_2.
Final $[H^+]$ = $[H^+$ released$]$ + $[H^+$ initial$]$ - $[HCO_3^-]$
= 0.352 + 1.07 x 10^{-7} - 1.64 x 10^{-3} = 0.350 mol/l

Final pH = 0.46
Therefore, the pH drop = 7 - 0.46 = 6.54

Case (II)
Pore diameter = 30 μm = 30 x 10^{-6} m
Concentration of H^+ released = 2.93 x 10^{-4} mol/l

At pH = 7.0 and $[HCO_3$-$]$ = 1.63 x 10^{-3} mol/l, from equation (7.A.1) for K_1 = 4.17 x 10^{-7},
Initial $[CO_2]$ = 3.67 x 10^{-4} mol/l

Final $[HCO_3^-]$ = 1.64 x 10^{-3} - 0.29 x 10^{-3} = 1.35 x 10^{-3} mol/l
Final $[CO_2]$ = 3.93 x 10^{-4} + 2.93 x 10^{-4} = 6.86 x 10^{-4} mol/l

From equation (7.A.1), $[H^+]$ = 2.13 x 10^{-7} mol/l
Final pH = 6.67
Therefore, the pH drop = 7 - 6.67 = 0.33

Similarly, the pH drop in the pores of different sizes with the iron(II) adsorption onto iron oxide coated sand at different pH were calculated (Fig. 7.16).

Chapter 8

SUMMARY AND CONCLUSIONS

8.1 INTRODUCTION

Iron is commonly present in groundwater worldwide. The presence of iron in the water supply is not harmful to human health, however it is undesirable. Bad taste, discoloration, staining, deposition in the distribution system leading to aftergrowth, and incidences of high turbidity are some of the aesthetic and operational problems associated with iron in water supplies. Iron removal from groundwater is, therefore, a major concern for water supply companies using groundwater sources. The WHO recommended guideline value of iron in drinking water is 0.3 mg/l and the EC directive has set a parametric value of 0.2 mg/l. In the Netherlands, the guideline value of iron in drinking water is ≤0.05 mg/l and several Dutch water supply companies are aiming at iron concentrations below 0.03 mg/l to minimise distribution network maintenance costs. Of the different methods available to control iron in water supplies (oxidation-filtration, ion exchange, lime softening, sub-surface iron removal, and sequestration), aeration followed by rapid sand filtration is the most commonly used.

Different mechanisms (physical, chemical, and biological) may contribute to the removal of iron in filters and the dominant mechanism depends on water quality and process conditions applied. Fig. 8.1 summarises the different mechanisms of iron removal in filters and the steps involved. Under anoxic conditions, adsorption is the only mechanism of iron removal from groundwater. In the presence of oxygen, iron removal can take place via three different mechanisms, namely i) oxidation-floc formation (floc filtration), ii) biological oxidation, and iii) adsorption-oxidation (adsorptive filtration). Under the commonly applied treatment conditions in iron removal plants, the oxidation-floc formation mechanism is commonly believed to be dominant. The adsorption-oxidation mechanism (adsorptive iron removal), however, has several potential advantages over the oxidation-floc formation mechanism, namely longer filter run, shorter filter ripening time, and less backwash water use and sludge production.

8.1.1 Adsorptive iron removal in filters

The adsorption-oxidation mechanism is most likely responsible for the removal of iron entering the filter bed in iron(II) form in conventional iron removal filters. It is also most likely the dominant iron removal mechanism in dry filters and sub-surface iron removal. In practice, primarily adsorptive iron removal in filters can be realised in two operational modes:

(a) *Intermittent regeneration mode:* In this mode filters are operated under anoxic conditions and the oxidation of iron(II) is consequently suppressed by avoiding aeration. On exhaustion of the iron(II) adsorption capacity of the filter media, the anoxic bed requires regeneration of the adsorption sites by oxidation of adsorbed iron(II). This can be accomplished by e.g. backwashing the filter with oxygen-rich water or with a chemical oxidant e.g. $KMnO_4$.

(b) *Continuous regeneration mode:* In this mode filters are operated under aerobic conditions to allow continuous regeneration of the exhausted adsorption sites. A low concentration of oxygen and/or a short pre-oxidation time is required to avoid the formation of iron hydroxide flocs. In this mode there are three possible options: a) dry filter, b) normal rapid filter operated at a high filtration rate and low height of supernatant, and c) normal rapid filter with low oxygen concentration (1-2 mg/l) in the feed water. Some iron flocs, however, will also be formed under these conditions and therefore backwashing will be required when maximum head loss is reached.

Fig. 8.1 Iron removal mechanisms in filters

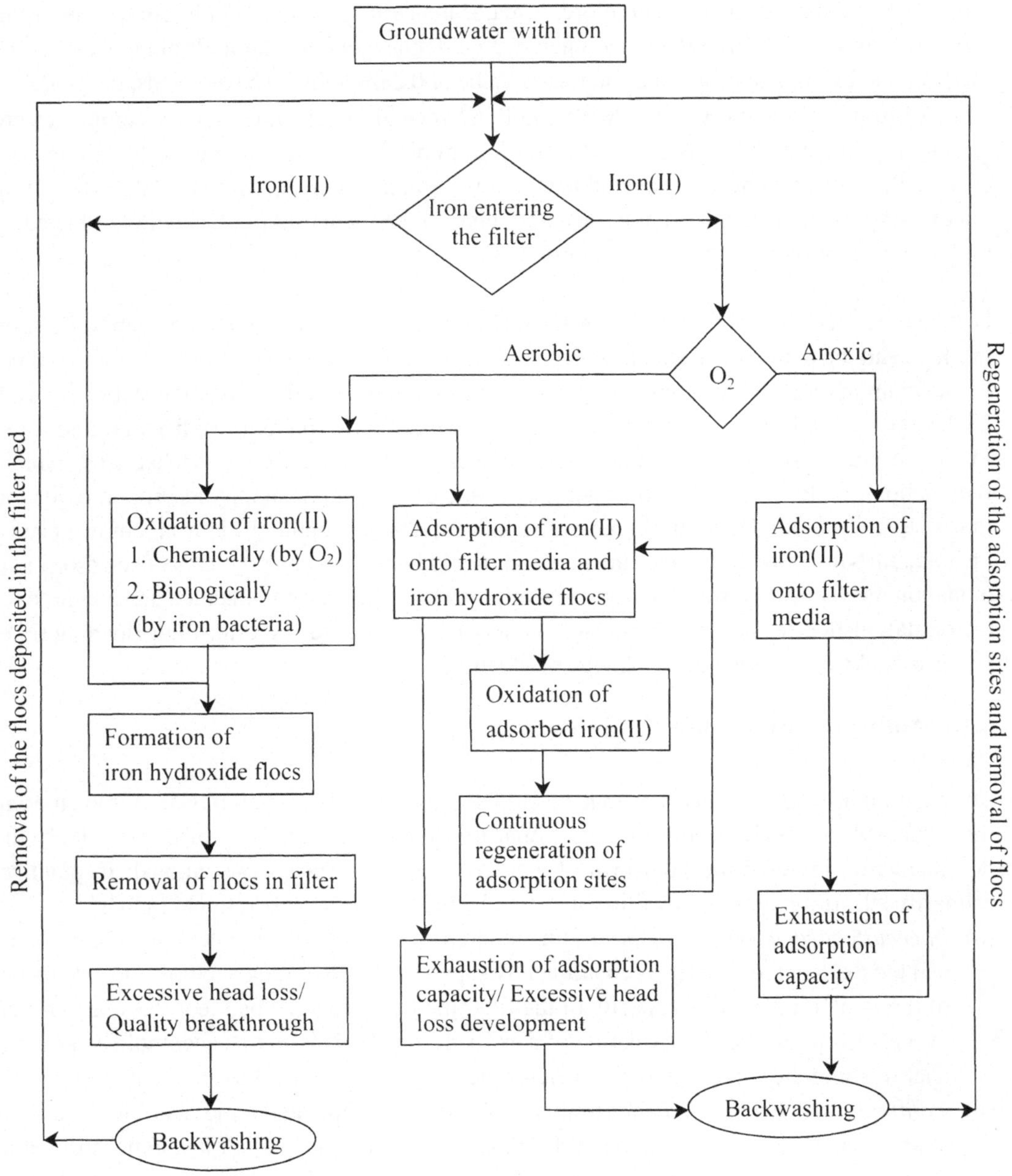

Potential advantages:

The adsorptive iron removal has the following potential advantages over floc filtration iron removal:

i) lower head loss development, hence longer filter runs as iron is adsorbed onto the surface of the filter media instead of forming flocs;

ii) shorter ripening time after backwashing;

iii) better average filtrate quality;

iv) operation and maintenance cost savings as less backwash water is used and there is no or less sludge to treat and dispose of. Additionally, the frequency of distribution system cleaning and consumer complaints would be reduced; however, excess iron oxide coated sand needs to be disposed of.

v) capital cost savings as the filter area and the backwash tank volume can be reduced by increasing the rate of filtration and decreasing the frequency of backwashing.

8.1.2 Need for research and objectives

Meeting the stringent requirements of iron in the drinking water supply (e.g. desired level of ≤0.03 mg/l in the Netherlands) and backwash water treatment in developed countries, and reducing the operation and maintenance costs of distribution systems worldwide will require a more efficient removal or minimisation of the iron currently passing the filters. Until now, it has been generally believed that, regardless of the water quality, the treatment approach was based on physical removal of the iron hydroxide flocs. However, recent publications reported that under "low oxygen conditions" biological iron removal takes place as well. Depending upon the water quality and process conditions applied, one mechanism may result in a higher process efficiency than another. To improve the efficiency of iron removal, it is therefore necessary that all the mechanisms involved are well understood and optimally utilised.

Adsorption-based processes are capable of removing metals over a wide pH range and to much lower levels than processes based on precipitation. Iron oxide coated sand has been successfully employed in fixed beds to treat different metal-bearing water and wastewater. However, no information is available on the effectiveness of the adsorption-oxidation mechanism when used as the dominant mechanism of iron removal from groundwater. Also, little is known about the effect of water quality parameters on iron(II) adsorption, the rate of oxidation of adsorbed iron(II), the process of development of iron oxide coating on filter media, and the characteristics of the iron oxide coating developed under different conditions. In order to get more insight into adsorptive iron removal and to further optimise the process of iron removal in filters, a detailed study of the iron(II) adsorption onto filter media was conducted. However, the factors governing the rate of oxidation of adsorbed iron(II) was not investigated in this study.

The goal of this research was to examine the potential of adsorptive iron removal as an alternative to the conventional floc formation method and to investigate the factors governing the adsorptive iron removal process, particularly the mechanisms involved under anoxic conditions.

This research had the following specific objectives:

i) To measure the iron(II) adsorption capacities of different new and iron oxide coated filter media and different iron (hydr)oxides, and to analyse the effect of pH on iron(II) adsorption capacity.

ii) To analyse the effect of major groundwater quality parameters namely Mn^{2+}, Ca^{2+}, NH_4^+, Cl^-, SO_4^{2-}, PO_4^{3-}, silica, alkalinity, and background ionic strength on iron(II) adsorption onto new and iron oxide coated sand.

iii) To analyse the physical and surface chemical characteristics of new sand and iron oxide coated sand developed under different water quality and process conditions.

iv) To study the factors affecting the development of iron oxide coating on the filter media.

v) To compare the performance of floc filtration and adsorptive filtration modes of iron removal by conducting pilot plant experiments.

vi) To model the process of adsorptive iron removal in filters and to predict the iron(II) breakthrough in adsorptive filtration columns with new and iron oxide coated sand at different pH under anoxic conditions.

8.2 ADSORPTION OF IRON(II) ONTO FILTER MEDIA AND IRON HYDROXIDES

Quantification of iron(II) adsorption capacities of filter media and iron hydroxides is essential for a better understanding and optimisation of adsorptive iron removal. Hence, a laboratory-scale batch experimental method was developed to measure the capacity of filter media and iron hydroxides to adsorb iron(II) ions. In addition to new (virgin) sand, iron oxide coated sand sampled from two groundwater treatment plants and other filter materials, namely anthracite, basalt, pumice, limestone, magnetite, and olivine were tested. Additionally, four iron (hydr)oxides namely lepidocrocite, goethite, amorphous iron hydroxide, and ferrihydrite were prepared in the laboratory and their iron(II) adsorption capacities were measured. All the batch adsorption experimental data fitted the Freundlich isotherm model. Based on the experimental results the following was concluded:

- Iron(II) adsorption capacities of filter media varied widely. Of the virgin materials tested, basalt showed the highest iron(II) adsorption capacity followed by anthracite, olivine, magnetite, sand, pumice, and limestone.
- Iron oxide coated sands from full-scale iron removal plants demonstrated a much higher capacity for iron(II) adsorption compared to new filter media. In the pH range examined (6-7.5), the iron(II) adsorption capacity of both new (virgin) and iron oxide coated sand increased with the increase of pH.
- The iron(II) adsorption capacities of coated sand measured in the model groundwater and in the real groundwater were different. This indicated that different ions and compounds present in groundwater could influence iron(II) adsorption.
- Among the iron hydroxides tested, lepidocrocite had the highest iron(II) adsorption capacity, followed by amorphous iron hydroxide, ferrihydrite and goethite. The iron(II) adsorption capacity of iron hydroxide also increased with the increase in pH. The iron(II) adsorption capacity of lepidocrocite decreased with ageing, probably due to its

transformation to other mineralogical forms of iron oxides.

- An estimation based on the experimental results indicated that the contribution of iron(II) adsorption onto iron hydroxide flocs on the overall process of adsorptive iron removal, as well as floc filtration iron removal, is likely negligible in iron removal plants.

8.3 EFFECT OF WATER QUALITY ON IRON(II) ADSORPTION

Different ions and compounds present in groundwater might influence iron(II) adsorption onto filter media and, hence, affect the process of adsorptive iron removal. Several laboratory-scale batch adsorption experiments were conducted to investigate the effect of major inorganic ions present in groundwater namely Mn^{2+}, Ca^{2+}, NH_4^+, SO_4^{2-}, Cl^-, PO_4^{3-}, silica, alkalinity, and background ionic strength on the iron(II) adsorption onto new silica sand and iron oxide coated sand. Based on the experimental results, the following conclusions were drawn:

- In the concentration range examined, NH_4^+ (0.25-1 mg/l), Cl^- (10-80 mg/l), alkalinity (1-8 mmol HCO_3^-/l), and background ionic strength (2.5-25 mmol $NaNO_3$/l), had no significant effect on iron(II) adsorption onto either new silica sand or iron oxide coated sand.
- Increasing concentrations of Mn^{2+}, Ca^{2+}, PO_4^{3-} and silica decreased the iron(II) adsorption onto new sand, whereas an increase in iron(II) adsorption was observed when SO_4^{2-} concentration was increased.
- Iron(II) adsorption onto iron oxide coated sand reduced with an increase in Ca^{2+} and silica concentration, however the effect of SO_4^{2-} and PO_4^{3-} was negligible.
- In general, it was observed that the effect of different ions and compounds on iron(II) adsorption is more pronounced on new silica sand than on iron oxide coated sand. Once the iron oxide coating is (fully) developed on the filter media, the adsorption of iron(II) is not hindered significantly by the presence of other inorganic ions within the concentration range common for groundwater.
- Preliminary experiments with commercial humic acid showed a negative effect of organic matter on iron(II) adsorption capacity. The effect of organic matter on the iron(II) adsorption capacity needs to be measured in groundwater from different locations since the nature of these organic compounds and their effect is most likely site specific.

8.4 CHARACTERISTICS OF COATED SAND FROM IRON REMOVAL PLANTS

Iron oxide coatings developed under different water quality and process conditions may have different physicochemical characteristics and, hence, different adsorption capacities. The physical and surface chemical characteristics of coated sand from twelve different groundwater treatment plants in the Netherlands were analysed to assess their potential use in adsorptive iron removal. The amounts of iron, manganese, calcium, and TOC in the coatings were determined by chemical extraction, and the iron(II) adsorption capacities of the coated sands were measured. Additionally, specific surface area, porosity, grain size distribution, and density of the media were measured and X-ray diffractograms, scanning electron micrographs, and EDAX spectra were prepared. The following conclusions were drawn:

- Compared to new sand, iron oxide coated sand had a very high porosity (up to 110 times) and a very large specific surface area (5-200 times).
- The higher the iron loading (time in use x raw water iron concentration), the higher was the surface extractable iron content of the coatings. The iron content of the coatings ranged from 27% to 45%, indicating a substantial difference in chemical composition of the coatings.
- At pH 6.5, the adsorption capacities of different coated sands were 10 to 55 times that of new sand. In general, iron(II) adsorption capacity was higher for the coated sand with a high surface extractable iron content and large specific surface area. However, the average annual increase of the iron content and the adsorption capacity varied for the coated sand from different plants, probably due to the difference in water quality, process conditions applied, and time in use.
- The grain size of the filter sand increased and the density decreased with the development of iron oxide coating. The decrease in density of coated sand with the iron oxide coating development was a function of the increase in the effective grain size.
- Iron oxides on the coatings were X-ray amorphous or poorly crystalline. SEM/EDAX analysis showed that iron and oxygen were the main components of the coating and elements like manganese, calcium, silicon, and carbon were also present in the coatings. It was observed that the coating developed on a filter grain was not uniform and different regions of the coating on a sand grain could have a different elemental composition and surface characteristics.
- The measured high adsorption capacities of coated sand from wet filters and dry filters of full-scale groundwater treatment plants indicate that, in wet filters, adsorptive iron removal also plays a role. In dry filters, this mechanism should be dominant due to a very short pre-oxidation time.
- As all the iron oxide coated sand tested demonstrated high iron(II) adsorption capacity, it is expected that process efficiencies of wet filters with coated sand can be further improved by operating the filters predominantly in the adsorptive filtration mode.

8.5 DEVELOPMENT OF IRON OXIDE COATING ON FILTER MEDIA

The development of an iron oxide coating on the filter media is an important factor in effective adsorptive iron removal from groundwater. The rate of development of the coating and its characteristics may be influenced by raw water quality, process conditions applied, and characteristics of the filter media. Laboratory-scale short column experiments were conducted to study the effect of pH, feed water iron concentration, and type of filter media on the development of iron oxide coatings on the filter media. The increase in the surface extractable iron content (SEIC) and iron(II) adsorption capacity of filter media with the coating development were measured. Based on the experimental results the following was concluded:

- The development of iron oxide coating on filter sand was faster at a pH of 6.5 and 7.0 than at a pH of 6.0. The coated sand developed at a pH of 6.5 and 7.0 had a considerably higher iron(II) adsorption capacity compared to coated sand developed at a pH of 6.0.
- Coating development was faster at feed water iron concentrations of 4.0 mg/l and 6.0 mg/l compared to 1.0 mg/l. The coated sand developed at the highest iron concentration tested (6 mg/l) had the highest iron(II) adsorption capacity. Initially, both the SEIC and the iron(II) adsorption capacity increased rapidly with the development of the coating. Further on, the adsorption capacity increased at a much slower rate despite a steady increase in the SEIC with time, most likely because the sand surface was fully covered with iron oxide and the newly deposited iron oxides covered and/or filled the pores in previously deposited layers.
- Among the three media tested, virgin basalt had the highest iron(II) adsorption capacity, and the coating development was fastest on basalt, followed by olivine and sand. The use of filter media with a high iron(II) adsorption capacity, like basalt, in iron removal can be an attractive option to reduce the time required to develop the coating and to improve the process efficiency.
- Preconditioning of new filter media at high feed water pH and/or high iron concentration is expected to be an attractive approach to rapidly develop an effective iron oxide coating and to reduce the initial filter ripening time. An alternative approach would be to circulate water with a high iron(II) concentration under anoxic conditions with regular backwashing with oxygen-rich water.
- Further studies should be conducted to analyse the effect of backwashing conditions on iron oxide coating development and to examine the effect of ageing of the coating on the iron(II) adsorption capacity.

8.6 COMPARISON OF PHYSICOCHEMICAL IRON REMOVAL MECHANISMS

It is expected that the process efficiency of iron removal in filters can be optimised with a better understanding of the different mechanisms involved in the process and their optimal utilisation. In order to investigate the options to improve the performance of an iron removal plant in the Netherlands, pilot experiments were conducted with filters operating in two different iron removal modes, namely floc filtration and adsorptive filtration. The results were then compared with the performance of the full-scale plant Gilze with sand media (0.8-1.25 mm), which operates mainly in the floc filtration mode. Results demonstrated that the performance of WTP Gilze in terms of ripening time, filter run time, and filtrate quality (residual iron and turbidity) could be improved by operating the filters a) mainly in adsorptive mode with fine sand (0.5-0.8 mm) and low influent oxygen, or b) mainly in floc filtration mode with dual media (anthracite 0.8-1.2 mm and sand 0.5-0.8 mm). The following was concluded:

- The filter with fine sand (0.5-0.8 mm) operating in floc filtration mode gave better filtrate quality and a much shorter ripening time than the full-scale plant. However, filter run times were unacceptably short.
- The single media filter with fine sand running mainly in adsorptive mode produced filtrate with low iron concentrations and had short ripening times. However, manganese and ammonium removal was not satisfactory due to low oxygen content in the feed water. In general, for the single media fine sand filter, the adsorptive mode gave a longer run time than the floc filtration mode. The filter run times of the adsorptive filter were, however, influenced by the influent oxygen concentrations. Filter run time criteria established for WTP Gilze (48 hours) was easily fulfilled at an influent oxygen concentration of 0.5-0.7 mg/l, while at an influent oxygen concentration of 1-2 mg/l, the filter run times were slightly shorter (40-44 hours).
- The use of dual media (anthracite 0.8-1.2 mm and sand 0.5-0.8 mm) considerably prolonged the run time of the filter operating mainly in floc filtration mode, and the filtrate quality satisfied all the evaluation criteria.
- Three options considered to be attractive to improve the performance of WTP Gilze are:
    - i) dual media filter in floc filtration mode.
    - ii) single media filter in adsorptive mode with coarser material (e.g. 0.7-1.2 mm) and low influent oxygen concentration (about 2 mg/l).
    - iii) dual media filter in adsorptive mode with low influent oxygen concentration (about 2 mg/l) and limited pre-oxidation time. This option needs to be verified in pilot or demonstration study.

  The experimental results suggest that the third option would result in the best performance in terms of both productivity and quality. However, for filters running under mainly adsorptive mode, the process conditions need to be selected carefully to ensure sufficient manganese and ammonium removal as well.

8.7 MODELLING ADSORPTIVE IRON REMOVAL FROM GROUNDWATER

Modelling of adsorptive iron removal is required to predict the performance of filters operated in adsorptive mode and to establish design parameters for adsorptive iron removal filters. Additionally, an accurate prediction of the breakthrough with a mathematical model can reduce the need for lengthy and costly pilot experiments. The process of adsorptive iron removal in filters under anoxic conditions only was modelled using adsorption isotherm parameters, mass balance and mass transfer equations. For this fixed bed adsorption model, the iron(II) breakthroughs in adsorptive filters with new and iron oxide coated sand were predicted using three different models with increasing complexity, and the model predictions were compared with the experimental results. The following conclusions were drawn:

- The experimental iron breakthrough in adsorptive filters occurred much earlier than the predictions of the Constant Pattern Model (CPM), which considers film (external) mass transfer only. In addition, the shapes of the predicted iron breakthrough curves were considerably different from the experimental breakthrough curves.
- The Linear Driving Force Model (LDFM), which considers an overall mass transfer combining both external and internal mass transfer, gave a good prediction for the initial breakthrough in filter columns with new sand. However, the subsequent development gave a significant deviation. For coated sand, the model predictions improved but were still quite different from the measured values.
- The Plug Flow Homogeneous Surface Diffusion Model (PFHSDM), which separately includes both film mass transfer and surface diffusion, gave a good prediction of initial breakthrough in filter columns with new sand. Further development of the curve, as in the case of the LDFM, gave a significant deviation from experimental results. For coated sand, the iron breakthrough predictions with PFHSDM were not better than with the LDFM.
- Incorporating dispersion in the LDF and the PFHSDM gave no substantial improvement in iron breakthrough predictions.
- The difference in model predictions and experimental results in the case of iron oxide coated sand was probably due to the effect of an initial pH drop in the pores with iron(II) adsorption, and a consequent decrease in iron(II) adsorption capacity.
- More accurate predictions of iron breakthrough will require quantification of the effect of pH change in the pores of the filter media with iron(II) adsorption. In addition, the effect of inhomogeneity of pore sizes, and traces of oxygen in the feed water and in the pores of the filter media needs to be studied in more detail.

8.8 GENERAL CONCLUSIONS

- *Iron oxide coated sand from iron removing filters has a high iron(II) adsorption capacity, which can be utilised, in principle, to improve iron removal in filters by switching the governing mode of operation from floc filtration to adsorptive filtration. This can be achieved by bypassing the aeration step and/or reducing the pre-oxidation time to ensure that the majority of the iron enters the filter bed in iron(II) form. In dry filters the dominant mechanism is already adsorptive iron removal.*

- *Iron(II) adsorption onto iron oxide coated filter media is not hindered significantly by the presence of other inorganic ions within the concentration range common for groundwater. The effect of organic matter present in groundwater on iron(II) adsorption needs to be studied in detail.*

- *Preconditioning of new filter media (e.g. at high feed water pH and/or high iron concentration) will result in rapid development of an effective iron oxide coating that will reduce the initial filter ripening time. Additionally, the use of virgin media with a high iron(II) adsorption capacity, like basalt, will also reduce the time required to develop a coating with adequate adsorption capacity.*

- *Adsorptive filtration can, in principle, be employed as the primary method of iron removal from anoxic groundwater without ammonium.*

- *Adsorptive iron removal could be very attractive when two filtration steps are applied due to high concentrations of iron, manganese, and ammonium in raw water. The first filter can be optimised as an adsorptive iron removal filter, while the second filter can be employed for manganese and ammonium removal.*

- *Adsorptive iron removal is potentially an attractive alternative to conventional floc filtration iron removal. Application of this process has prospects of improving the filtrate quality, extending the filter run time, and easing the treatment of filter backwash water and sludge, thus resulting in a higher treatment process efficiency.*

8.9 RECOMMENDATIONS FOR FUTURE RESEARCH

1. The efficiency and continuity of the adsorptive iron removal process depends on the rate of regeneration of adsorption sites or the rate of oxidation of absorbed iron(II). Therefore, it is necessary to analyse factors influencing the rate of oxidation of adsorbed iron(II), with special attention to the following:
    - water quality parameters (pH, O_2, inorganic compounds, organic matter) and
    - filter media (new /iron oxide coated)

   In this context, it is necessary to model the rate of oxidation of adsorbed iron(II) in filters, considering the external and internal mass transfers of O_2 and the rate of filtration or backwashing.

2. Filter media of adsorptive iron removal filters are expected to grow with time. The effect of water quality and process conditions on the rate of growth of filter media needs a further detailed investigation. Additionally, further studies should be conducted to analyse the effect of backwashing conditions on iron oxide coating development and to examine the effect of ageing of the iron oxide coating on the long term iron(II) adsorption capacity.

3. Models for the prediction of iron(II) breakthrough in adsorptive filters operated under intermittent regeneration mode should be further improved, considering the effect of pH drops in pores of the filter media with iron(II) adsorption, inhomogeneity of pore sizes, and traces of oxygen in the feed water and in the pores of the filter media. Furthermore, mathematical models should be developed to predict the breakthrough of iron(II) in adsorptive filters operated under continuous regeneration mode.

4. Iron and manganese are normally present together in groundwater. This study concentrated on adsorptive iron removal only. Hence, a detailed investigation of the factors influencing the adsorptive manganese removal from groundwater is required.

5. Iron oxide coated sand has the capacity to adsorb a host of cations and anions. Studies at IHE have demonstrated the high potential of iron oxide coated sand in arsenic removal. Therefore, a further study on the combined removal of arsenic, iron, and manganese from groundwater in the adsorptive filtration mode is recommended.

Samenvatting

IJzer is veelvuldig aanwezig in grondwater. In zuurstofloos grondwater is ijzer meestal in opgeloste, tweewaardige vorm aanwezig. IJzer in drinkwater is niet schadelijk voor de gezondheid, de aanwezigheid wordt evenwel als ongewenst beschouwd. Smaak, kleur, vlekken op kleding, afzettingen in het distributiesysteem resulterend in nagroei en regelmatig hoge troebeling, zijn de voornaamste bezwaren, die met ijzer in drinkwater worden geassocieerd en daarom alle aandacht van de grondwaterbedrijven heeft.

Beluchting gevolgd door snelfiltratie is de meest toegepaste methode voor de verwijdering van het opgeloste ijzer uit grondwater. Verschillende mechanismen (fysisch, chemisch en biologisch) dragen bij aan de ontijzering in snelfilters. Welk mechanisme domineert is afhankelijk van de waterkwaliteit en de bedrijfvoering. Algemeen wordt aangenomen dat bij de ontijzering van grondwater onder de gebruikelijke omstandigheden het mechanisme van "oxidatie gevolgd door vlokvorming en vlokverwijdering door filters" (vlokkenfiltratie) domineert. Hierbij wordt het opgeloste ijzer(II) dat in zuurstofloos grondwater aanwezig is geoxideerd tot ijzer(III) en - na precipitatie - als ijzerhydroxidevlokken verwijderd door snelfilters. Als mogelijk tweede ontijzeringsmechanisme kan "adsorptie van ijzer(II) aan het filtermateriaal, gevolgd door oxidatie" (adsorptieve filtratie) worden genoemd. In dit proces wordt ijzer(II) verwijderd door adsorptie aan het oppervlak van het filtermateriaal waarna het geadsorbeerde ijzer(II) wordt geoxideerd, resulterend in een nieuw adsorptie-oppervlak. Adsorptieve filtratie heeft ten opzichte van vlokkenfiltratie enkele potentiële voordelen zoals een langere filterlooptijd, betere kwaliteit, een kortere rijpingstijd en een geringere produktie van spoelwater en slib. Bij conventionele ontijzeringsfilters is te verwachten dat het proces van adsorptie gevolgd door oxidatie verantwoordelijk is voor de verwijdering van een deel van het in de voeding van de filters aanwezige ijzer(II). Dit mechanisme is waarschijnlijk dominant bij droogfiltratie en ondergrondse ontijzering.

Waterleidingbedrijven streven naar de toepassing van mogelijkheden die de ontijzering van grondwater kunnen verbeteren om de afzetting van ijzer in het distributienet, het spoelwaterverbruik en de slibproduktie te minimaliseren. De WHO-richtlijn voor het ijzergehalte in drinkwater is 0,3 mg/l; de richtlijn van de Europese Gemeenschap is 0,2 mg/l. In Nederland streven verschillende waterleidingbedrijven naar waarden gelijk of lager dan 0,03 mg/l. Deze stringente eisen in dit opzicht, de noodzaak van spoelwaterbehandeling in westerse landen en de reductie van de bedrijfskosten van distributiesystemen vereisen een doelmatiger verwijdering van ijzer en/of minimalisatie van ijzerdoorslag in de filters. Tot nu toe wordt er ongeacht de waterkwaliteit van het grondwater van uit gegaan dat de ontijzering is gebaseerd op vlokkenfiltratie. Afhankelijkheid van de waterkwaliteit en de processcondities kan echter de toepassing van adsorptieve filtratie in principe een beter resultaat geven dan vlokkenfiltratie. Een beter begrip van de verschillende mechanismen die bij de ontijzering een rol spelen is van wezenlijk belang voor een optimaal ontwerp en bedrijfsvoering van ontijzeringsfilters. Ondanks de potentiële voordelen wordt tot op heden adsorptieve filtratie nog niet bewust toegepast in snelfilters.

De doelstelling van het onderzoek in dit proefschrift is de mogelijkheden te onderzoeken van adsorptieve ontijzering als alternatief voor de conventionele vlokkenfiltratie en na te gaan welke factoren het adsorptieve ontijzeringsproces met name in zuurstofloos milieu bepalen. Dit onderzoek is gedaan door i) het meten van de adsorptiecapaciteit voor ijzer(II) van enkele nieuwe filtermaterialen, van met ijzeroxide gecoat zand van ontijzeringsfilters en van ijzerhydroxide, ii) het effect van waterkwaliteitsparameters op de adsorptiecapaciteit te meten, iii) de invloed van pH, ijzerconcentratie en het type filtermateriaal op de ontwikkeling van de ijzeroxidecoating te bepalen, iv) de werking van proeffilters uitgevoerd als adsorptieve filters en vlokkenfilters te vergelijken en v) de adsorptieve ontijzering te modelleren in filters teneinde de ijzerdoorbraak in filters - voorzien van nieuw en met ijzeroxide gecoat filtermateriaal - in zuurstofloos milieu te kunnen voorspellen.

Een methode is ontwikkeld voor de meting van de adsorptie van ijzer(II) aan filtermaterialen en ijzerhydroxide. Onderzoek wees uit dat het adsorptieproces van ijzer(II) aan filtermaterialen zowel met de Freundlich- als met de Langmuir-isotherm kan worden beschreven. De adsorptiecapaciteit van de verschillende filtermaterialen verschilden hierbij sterk. Van de onbehandelde filtermaterialen leverde basalt de hoogste adsorptiecapaciteit gevolgd door anthraciet, olivijn, magnetiet, zand, puimsteen en kalksteen. Met ijzeroxide gecoat zand afkomstig van ontijzeringsfilters van grondwaterpompstations vertoonde een opmerkelijk hogere adsorptiecapaciteit dan nieuw zand. In het onderzochte pH-gebied (6-7,5) leidde een verhoging van de pH voor zowel het met ijzeroxide gecoat zand als nieuw zand tot een toename van de adsorptiecapaciteit voor ijzer(II). Van de onderzochte ijzerhydroxiden vertoonde lepidocrociet de hoogste adsorptiecapaciteit, gevolgd door amorf ijzerhydroxide, ferrihydriet en goethiet. Uit het onderzoek is af te leiden dat in ontijzeringsfilters de bijdrage van adsorptie van ijzer(II) op ijzerhydroxidevlokken aan het adsorptieve ontijzeringsproces in zijn geheel, als ook aan de op vlokkenfiltratie gebaseerde ontijzering te verwaarlozen is.

De hoge adsorptiecapaciteit voor ijzer(II) van met ijzeroxide gecoat zand afkomstig van grondwaterpompstations kan in principe worden benut voor de verbetering van de ontijzering, door om te schakelen van vlokkenfiltratie naar adsorptieve filtratie. Om zeker te stellen dat het merendeel van het ijzer in tweewaardige vorm het filter bereikt, kan de beluchtingstrap gedeeltelijk worden kortgesloten en/of de vooroxidatietijd worden beperkt. In de praktijk kan adsorptieve filtratie voornamelijk op twee wijzen worden gerealiseerd, te weten a) een periodieke regeneratiewijze door het filter in afwezigheid van zuurstof te bedrijven en de adsorptieplaatsen te regenereren door bijvoorbeeld het filter terug te spoelen met zuurstofrijk water of te behandelen met een oxidatiemiddel zoals $KMnO_4$ en b) middels een continue regeneratiewijze door de filters in aanwezigheid van een beperkte hoeveelheid zuurstof te bedrijven en/of door de vooroxidatietijd te beperken.

Binnen het onderzochte concentratiegebied bleken NH_4^+, Cl^-, HCO_3^- en ionensterkte geen duidelijke invloed te hebben op de ijzer(II)-adsorptie noch op nieuw kwartszand noch op met ijzeroxide gecoat zand. Een verhoging van de concentratie van Mn^{2+}, Ca^{2+}, PO_4^{3-} en SiO_2 leidde tot een verlaging van de ijzer(II) adsorptie op nieuw zand terwijl daarentegen een verhoging van de SO_4^{2-}-concentratie een adsorptieversterkende effect had. Op het gedrag van met ijzer gecoat zand hadden voornoemde ionen vrijwel geen invloed. In het algemeen was het effect van verschillende parameters op de ijzer(II)-adsorptie meer uitgesproken voor nieuw zand dan voor met ijzeroxide gecoat zand. Dit betekent dat waneer de ijzeroxidecoating zich eenmaal gevormd heeft op het oppervlak van het filtermateriaal de ijzer(II)-adsorptie binnen het gangbare concentratiegebied voor grondwater niet noemenswaard wordt gehinderd door de aanwezigheid van andere anorganische ionen. In orienterende experimenten met commercieel verkrijgbaar humuszuur vertoonde dit een negatief effect op de adsorptiecapaciteit. Het effect van organische stof, zoals het in grondwater voorkomt dient nader te worden bestudeerd.

Uit de analyse van de fysische en chemische eigenschappen van het oppervlak van gecoat zand van twaalf grondwaterpompstations in Nederland bleek dat - vergeleken met nieuw zand - gecoat zand een hoge porositeit en een groot specifiek oppervlak heeft. In het algemeen nam het ijzergehalte van de coating en de ijzer(II)-adsorptiecapaciteit bij gebruik toe in de tijd. De gemiddelde jaarlijkse toename van het ijzergehalte en de adsorptiecapaciteit wisselden voor de verschillende bedrijven, waarschijnlijk ten gevolge van het verschil in waterkwaliteit, procesomstandigheden en gebruiksduur. De optredende dichtheidsreduktie van het gecoate zand was - tijdens de ontwikkeling van de coating - gerelateerd aan de toename van de effectieve korreldiameter. De gemeten hoge adsorptiecapaciteiten van gecoat zand afkomstig van droog- en natfilters van grondwaterpompstations wijzen uit dat bij natfilters adsorptieve ijzerverwijdering ook een rol speelt. Bij droogfilters zou dit mechanisme kunnen domineren vanwege de korte vooroxidatietijd.

De ontwikkeling van een ijzeroxidecoating op de filtermaterialen is een belangrijke stap voor een doelmatige ontijzering van grondwater. De groeisnelheid van de coating en zijn eigenschappen zijn afhankelijk van de ruwwaterkwaliteit, de procesomstandigheden en de eigenschappen van de filtermaterialen. Het bleek dat conditionering van nieuwe filtermaterialen zoals toepassing van een hoge pH en/of hoge ijzerconcentratie tot een effectieve coating leidde. Dit kan in de praktijk een reductie van de filterrijpingstijd betekenen. Toepassing van nieuwe filtermaterialen met hoge ijzer(II)adsorptiecapaciteit zoals basalt een extra bijdrage leveren aan de reductie van de tijd nodig om een coating met afdoende adsorptiecapaciteit te ontwikkelen.

Het adsorptieve ontijzeringsproces in filters in zuurstofloos milieu is gemodelleerd met behulp van adsorptie-isotherm parameters, massabalanzen en stofoverdrachts- vergelijkingen. Experimentele resultaten zijn vergeleken met de voorspellingen van drie adsorptiemodellen, te weten i) het Constant Pattern Model (CPM), ii) het Linear Driving Force Model (LDFM) en iii) het Plug Flow Homogeneous Surface Diffusion Model (PFHSDM). Het CPM-model, dat alleen de uitwendige diffusie in beschouwing neemt, beschrijft de doorbraak van ijzer in filters met nieuw en met ijzeroxide gecoat zand onvoldende. Bij toepassing van het LDFM en het PFHSDM model blijkt dat doorbraak van ijzer beter te voorspellen is voor nieuw zand. Voor met ijzeroxide gecoat zand zijn de voorspellingen echter onbevredigend. Het verschil tussen de voorspellingen en de gemeten waarden voor gecoat zand is waarschijnlijk te wijten aan het optreden van een pH daling in de porieën van het filtermateriaal met als gevolg een aanzienlijk lagere adsorptiecapaciteit.

Adsorptieve filtratie kan als dominant mechanisme voor de ontijzering van zuurstofloos grondwater worden toegepast althans wanneer mangaan en ammonium afwezig zijn. Dit proces zou ook aantrekkelijk kunnen zijn voor toepassing in situaties waar een tweetrapsfiltratie wordt toegepast in verband met hoge ijzer-, mangaan- en ammoniumconcentraties in het grondwater. Het eerste filter kan optimaal benut worden voor de adsorptieve ontijzering, terwijl het tweede filter de ontmanganing en de nitrificatie verzorgt.

Adsorptieve ontijzering is potentieel een aantrekkelijk alternatief voor ontijzering gebaseerd op vlokkenfiltratie. Toepassing van dit proces geeft uitzicht op een verbetering van de filtraatkwaliteit, verlenging van de looptijd, verlichting of beperking van de spoelwaterbehandeling en de slibverwerking, aldus leidend tot een hoger zuiveringsrendement.

Trefwoorden: grondwater, ontijzering, verwijderingsmechanismen, vlokvorming, adsorptie, proces-efficiency, adsorptiecapaciteit, met ijzeroxide gecoat zand, waterkwaliteit, coating, modellering, stofoverdracht.

List of publications

Publications in Journals

1. Sharma, S.K., Greetham, M.R. and Schippers, J.C. (1999) Adsorption of iron(II) onto filter media. *Journal of Water Supply: Research and Technology - Aqua*, **48** (3), 84-91.

2. Sharma, S.K., Mendis, B.S. Greetham, M.R. and Schippers, J.C. (2000) Modelling adsorptive iron removal in filters. *Water Supply*, **18** (1-2), 604-608.

3. Sharma, S.K., IJpelaar, G. and Schippers, J.C. (2001) Iron oxide coating development on filter media. *Transactions of The Filtration Society*, **1** (3), 68-71.

4. Sharma, S.K., Kappelhof, J., Groenendijk, M. and Schippers, J.C. (2001) Comparison of physicochemical iron removal mechanisms in filters. *Journal of Water Supply: Research and Technology - Aqua*, **50** (4), 187-198.

5. Sharma, S.K., Sebwato, C., Petrusevski, B. and Schippers, J.C. (2001) Effect of groundwater quality on adsorptive iron removal. Accepted for publication in *Journal of Water Supply: Research and Technology - Aqua.*

6. Sharma, S.K., Petrusevski, B., Heijman, B. and Schippers, J.C. (2001) Prediction of iron(II) breakthrough in adsorptive filters: Intermittent regeneration mode. *Journal of Water Supply: Research and Technology - Aqua* (In preparation).

Publications in Conference Proceedings

7. Sharma, S.K., Petrusevski, B. and Schippers, J.C. (2000) Modelling adsorptive iron removal from groundwater. *Proceedings of IWA Workshop on Modelling of Conventional Drinking Water Production Processes (16-18 March 2000), Delft, The Netherlands.*

8. Sharma, S.K., IJpelaar, G. and Schippers, J.C. (2000) Development of iron oxide coating on filter sand. *Proceedings of 8th World Filtration Congress (3-7 April 2000), Brighton United Kingdom,* **2**, 853-856.

9. Sharma, S.K., Sebwato, C., Petrusevski, B. and Schippers, J.C. (2000) Effect of groundwater quality on adsorptive iron removal. *Proceedings of IWA Conference on Innovations in Classical and Conventional Water Treatment Processes (26-29 September 2000), Amsterdam, The Netherlands*, 15, 1-12.

10. Sharma, S.K., Petrusevski, B. and Schippers, J.C. (2001) Adsorptive iron removal: An alternative to the conventional approach. *Proceedings of Asian Waterqual 2001: First IWA Asia-Pacific Regional Conference (12-15 September 2001), Fukuoka, Japan*, **2**, 3-8.

11. Sharma, S.K., Petrusevski, B. and Schippers, J.C. (2001) Characterisation of coated sand from iron removal plants: *Proceedings of 2nd IWA World Water Congress (15-19 October 2001), Berlin, Germany.*

Curriculum vitae

Saroj Kumar Sharma was born in Kathmandu, Nepal on 19 August 1963. He received his B.E. (Civil Engineering) degree with honours in 1988 from Malaviya Regional Engineering College, University of Rajasthan, India.

After graduation, he joined His Majesty's Government of Nepal and worked for the Department of Irrigation and the Department of Water Supply and Sewerage in various capacities. He has wide experience in the water supply and sanitation sector ranging from planning, design, implementation, and operation and maintenance of community-based rural, semi-urban, and urban water supply and sanitation projects to human resources development and management information system. He has also worked with several national and international consultants and donor agencies in water supply and sanitation projects in Nepal, Bhutan, and Pakistan.

He came to IHE, Delft in 1995 and obtained the post-graduate diploma in Sanitary Engineering with distinction in 1996 and an M.Sc. in Sanitary Engineering with distinction in 1997. After that, he continued working as a Ph.D. research fellow in a joint research project of IHE Delft, Kiwa N.V. Water Research, and five water supply companies of the Netherlands namely, Water Supply North West Brabant (WNWB), Water Supply East Brabant (WOB), Water Supply Gelderland (WG), Water Supply Drenthe (WMD) and Water Supply Overijssel (WMO).